The Gut Chronicles

For everyone who has felt lost and struggled in silence, while navigating a troublesome gut, I see you.

For my father… It took me some years but here it is…

The Gut Chronicles

Sandra Mikhail APD

Foreword by
Maree Ferguson
Director, Dietitian Connection

Hammersmith Health Books
London, UK

First published in 2023 by Hammersmith Health Books
– an imprint of Hammersmith Books Limited
4/4A Bloomsbury Square, London WC1A 2RP, UK
www.hammersmithbooks.co.uk

British Library Cataloguing in Publication Data: A CIP record of this book is available from the British Library.

Print ISBN 978-1-78161-229-3
Ebook ISBN 978-1-78161-230-9

Commissioning editor: Georgina Bentliff
Typeset by: Julie Bennett of Bespoke Publishing, UK
Cover design by: Madeline Meckiffe based on a concept by Amélie Buri
Cover image by: Amélie Buri
Cartoons by: Amélie Buri
Index: Dr Laurence Errington
Production: Deborah Wehner of Moatvale Press Ltd
Printed and bound by: TJ Books Limited, Cornwall, UK

Contents

Contents

Contents

Contents

Foreword

I am honored to write the foreword for *The Gut Chronicles* and introduce readers to author, Sandra Mikhail. I have known Sandra for many years, and her passion and expertise for all things gut health are exceptional.

Good health starts in the gut, but many adults are yet to reach their gut-health potential, with at least 40% of people experiencing uncomfortable gut symptoms like bloating and diarrhoea. As Sandra outlines, new research is uncovering just how central the gut is to all aspects of wellbeing, including the immune system and mental health. As such, there has never been a better time to focus on the health of your insides.

There is nobody more qualified to guide you to good gut wellbeing than Sandra. Sandra has long suffered from debilitating gut health issues herself, she is a trained dietitian and nutritionist and her father is a gastroenterologist (also known as a gut doctor). I have witnessed first-hand the passion Sandra has for improving the gut health of her patients, and I know this labour of love, *The Gut Chronicles*, will enhance the gut health of anyone who reads it. It is an excellent resource for those suffering from gut health concerns, as well as dietitians, doctors, nurses and other allied health professionals. It is a comprehensive, up to date guide on the most common gut disorders and filled with practical tips and tricks to help people live and eat with confidence.

An absolute must-have on your bookshelf!

Maree Ferguson
Director, Dietitian Connection

About the Author

Sandra Mikhail is an internationally-known and accredited, practising dietitian and the founder and director of Nutrition A-Z by Sandra Mikhail. She holds a Bachelor of Nutrition and Dietetics (Monash University, Australia) and a Master of Advanced Studies in Nutrition and Health (ETHZ), and is a member of Dietitians Australia. She also holds a Sports Nutrition Diploma from the International Olympic Committee (IOC).

With experience spanning well over a decade, Sandra's main areas of specialty are digestive disease, sports nutrition and eating disorders. Her passion for and work in gut health created a movement for normalising 'poo talk', shedding light on topics that you may find yourself secretly searching the internet for. Known as the gut health dietitian 'making poo talk salon chic', she currently lives in Zürich, Switzerland, with her husband and children, with a thriving private practice and a signature tea blend for turbulent tummies.

1

Introduction

The long road home involved driving on a stress-inducing freeway where drivers have zero-tolerance for speed limits and tailgating is a regular occurrence. I still had about 10 minutes left until I was able to grace the porcelain throne, but my gut had other plans.

As I drove up to the final set of lights, I broke into a sweat, my stomach cramps felt like an earthquake rumbling within my lower extremities and I braced myself for an apocalypse. Anxiously staring at the lights, I looked to my left and noticed a young, flirtatious-looking driver locking eyes and cracking a smile at me without a hint in the world that the woman next to him was about to experience a boo-boo in her pants. 'Yep, it's happening,' I thought to myself as I awkwardly smiled back.... Yes, it happened. The lights turned green, foot on pedal and off I drove back home with the most uncomfortable and revolting feeling of having pooed my pants, in my car, as I smiled back at a stranger.

My passion for gut health was definitely born of a mix of being the daughter of a gastroenterologist and experiencing the unpredictable and painful symptoms of IBS. This three-letter acronym stands for 'irritable bowel syndrome' and, for those who are still not familiar with the term, it is a glitch in something we'll be covering in detail called the 'gut-brain axis', causing a disturbance in the functioning of the digestive tract (i.e. intestines). Bloating, gas, irregular bowel movements, abdominal pain and urgency to use the loo are some common symptoms many sufferers describe (ah, the joy). Some people with IBS complain of constipation, others have diarrhoea and some experience both.

To cut a story spanning over a year short, I started experiencing symptoms on and off after suffering a horrible stomach infection. That also happened to coincide with my moving from Australia to Dubai. So, change of environment? Food? Weather? Who knows. Following a course (and mix) of antibiotics, my guts of steel transformed into the weakest link. My symptoms could only be described as an abdominal roller-coaster: stomach pain then diarrhoea for a few days followed by no intestinal movement what-soever. Everything I ate seemed to go straight through me or create the perfect pregnancy illusion, where feeling bloated was simply an understatement. After endless trips to the bathroom (and moments when I'd literally lost control of my bowels, such as my poo-in-car story), the investigations began. Blood work, check. Endoscopy, check. Poo samples (many, in all forms), check. The tests ruled out coeliac disease, inflammatory bowel disease and other not so great diagnoses, and eventually, it all came down to IBS and lactose intol-erance with a lovely dash of reflux.

Being a dietitian, you would think 'Hold on, check for intoler-ances? Sensitivities? Try an elimination diet? C'mon!'. However, due to my insane fatigue and endless trips to the oval office, I thought something was seriously wrong that went far beyond that. However, 12 years post IBS diagnosis, I found myself embarking on a low-FOD-MAP diet, adding probiotics to the mix and embracing ways to manage stress. There was a light at the end of that shitty tunnel!

The reason for sharing a little snippet of my story is to break the taboo around poo talk and to reassure you that digestive prob-lems, no matter how mild or severe, need to be addressed out loud. Growing up, our dinner table talk involved a lot of hospital calls my dad would have to answer, shouting out things like 'rectal bleeding', 'anal fissures' and, yes, lots more dinner-inappropriate terminol-ogy. The horror on our friends' faces whenever they visited was hysterical to say the least.

During the course of my career, I have been very vocal about topics that you may find yourself secretly searching the internet for. This book will definitely include a lot of faecal *tête-à-têtes* so if you're squeamish and lacking any tolerance for potty-talk, I apologise in advance as you're not getting a refund, but most importantly, this book also offers a fad-free look at how to manage some of these con-ditions through diet and lifestyle. Throughout this journey together, we'll be going through the most common gut conditions that we,

homo sapiens, experience, focusing on the nutritional management of each.

Now, gut health has become completely mainstream, and pseudoscience continues to inundate people's minds with false claims,
promises and celery talk, making any responsible health professional
cringe and bang their head in disappointment. Luckily, there will be
no excessive celery talk in this book. I have created my brand with
a mission to help people navigate the complex world of health and
wellness using science as a backbone but, most importantly, using
easy-to-digest language that won't, I hope, lose you.

With this book, I want you to appreciate the complex nature of
your gut but also to use what I have written as guidance if you're
struggling. This comes from someone who has both the qualifications and bowel-changing experience.

Personalised nutrition

A question you may ask is, how can I personalise this general information? Before we get to that answer, let me grab your attention for
a minute to talk about the whole concept of personalised nutrition.

Nutrition, as we know, is an essential pillar of health and, as
a dietitian, personalised nutrition forms the basis of my practice.
Within this profession, (although everyone, their grandmothers
and Marie the neighbour have become experts in the field), personalised nutrition uses a whole heap of checks to create your lifestyle
template towards better health. These include your medical history,
blood work, sleep, stress and activity levels, emotions, thoughts and
behaviours around food and, in my line of work, good ol' poo. Yep,
your bowel movements and stool consistency can say a lot about
your overall health. You should see my clients' faces when I ask
them, in detail, about their bowel habits. Based on all the information collected, I formulate a nutritional plan that considers the
person as a whole. For example, I may advise George who has been

recently diagnosed with IBS to cut down on caffeine and go on a specific diet to manage his diarrhoea, OR I may ask Linda to include some kiwi fruit, prunes and flaxseed as well as more exercise to help her manage constipation. So, the evolution of personalised nutrition actually started with nutritionists and dietitians.

Personalised nutrition is now evolving with two major tools coming onto the scene – DNA testing and analysing your gut 'microbiome', a term we will dive into later. Analysing stool has always been a common diagnostic tool, in a lot of cases involving gut conditions, but more recently, we have been seeing the 'Christopher Columbus' of poo tests created to uncover a portal into the world of our gut microbiome. Microbiome testing and analysis are still at an early stage, but unfortunately (and as always) commercialisation is way ahead of the science. However, tools are slowly being developed, giving us insight into the types of little critters living in our gut, what foods they need, what their roles are, and how they are connected to other body organs and our overall health.

I won't be handing out free stool tests with this book but, for now, I want you to use the knowledge shared here not only to create your own template for managing any of the gut conditions mentioned but also to learn that there are specialised health professionals out there who can truly help you experience another level of personalised nutrition. And by health professionals, I'm talking about specialised dietitians and registered nutritionists, not social media wellness unicorns promoting a combination of celery juice, DIY enemas and other magical leprechaun juices that will cure you from all ailments.

A note about gut cancers

This book is about common, often chronic gut problems that are not life-threatening but have a significant negative impact on quality of life. I do therefore not address cancer; clearly 'gut' cancers warrant a whole book in themselves. Here I just want to offer a snapshot of what they are and warning signs to look out for, given how prevalent they are. Mutations (i.e. changes in our DNA) are known to be able to trigger the development of tumours anywhere along our digestive tract, leading to the development of the big C. The causes behind developing a type of gastrointestinal cancer are complex and multifactorial but it is a combination of genetics, environmental and lifestyle factors. The most common types include:

- Oesophageal cancer
- Liver cancer
- Stomach (gastric) cancer
- Colorectal cancer
- Pancreatic cancer

Colorectal cancer is the third most common cancer worldwide and is a type with which I have become far too familiar throughout my career.

If I had to pick a handful of essential points you need to bear in mind on the subject of gut cancers, they would include:

- Do not skip regular screening, especially if you have a strong family history of bowel cancer.

- Take action by seeing your doctor if you experience a combination of any of these **red flags**: blood in your poo; unexplained weight loss; unexplained anaemia; any bumps ('masses') along your abdomen, anus or rectum; unexplained diarrhoea; loss of appetite; vomiting blood.

- There is good news! Lifestyle changes discussed throughout this book can help in reducing your risk of developing gut cancers. These include reducing alcohol and red meat consumption, not smoking, moving regularly and having a diverse range of plants in your diet with a focus on fibre.

So bear in mind throughout this book that every time you follow the lifestyle advice that helps any gut problem, you will also be reducing your risk of cancer.

2

The one that takes you back to basics

The first question I want to address is simple: What actually is gut health?

Gut health encompasses a beautifully constructed world that works in synergy, influencing our overall wellbeing. Ancient thinking placed a focus on our gut's role in digestion, but the last 10 years have uncovered layer upon layer of detail in understanding our gut's personality. First and foremost, let's cover some basic physiology of digestion, something that all my readers need to grasp, especially when it comes to understanding common digestive disorders.

As we all know, food enters from one end and what we can't digest exits from another. For those who thought that digestion starts in your stomach, you were wrong. My friends, it all starts in your mouth! Let's take a (hopefully uncomplicated) trip through the gut, guided by Figure 2 on the next page.

Our whole digestive tract measures around 9 metres long. The components are:

Mouth: It is the beginning of the digestive tract and digestion starts as soon as you take the first bite of food. Chewing breaks down the food into smaller pieces that are easier to digest and the saliva mixes with the food to begin the process of breaking it

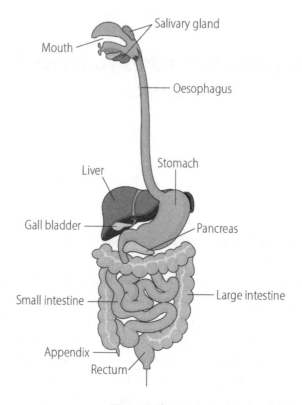

Figure 2: The gut

down. Your saliva contains compounds called 'enzymes' (principally 'amylase') that are responsible for breaking down food.

Oesophagus: The food travels from your mouth (after being pushed back into your throat by your tongue) through your throat and then into your oesophagus, which delivers the food into your stomach. This long tube that connects your mouth to your stomach is lined with muscles that contract, helping your food to move down via an action called peristalsis. The oesophagus also ends with what is called the 'oesophageal sphincter', a circle of muscle that when constricted prevents food from passing backwards from the stomach into the oesophagus.

Stomach: Consider this as the mixer and the grinder; your stomach secretes acid and powerful enzymes that continue the process of breaking down food. When the food leaves your stomach, it should be the consistency of a liquid or paste.

Small intestine: I find it comical that we call it 'small' since it's the largest part of our digestive system. Measuring around six metres long, your small intestine awaits the arrival of the ground-up food, which can take anywhere between one and four hours after eating. This long tube loosely coiled in the abdomen, continues to break down the food by using enzymes released by the pancreas and bile from the liver. Your small intestine is the primary site of nutrient absorption.

Large intestine: Also known as your colon, measuring less than the small intestine at 1.5 metres long, the large intestine moves along leftover waste from the digestive process (i.e. poo) by using contractions. It removes water from the stool, making it relatively solid, and empties into the rectum once or twice a day. (Fun fact: undigested food can actually take anywhere between 12 and 30 hours to move through your large intestine.) This part of our body also houses the trillion of gut bugs that make up the wonderful world of our gut microbiome, to which I've dedicated a whole chapter (see Chapter 3).

Rectum: The chamber that connects the large intestine to the anus, the rectum lets you know when stool needs to be evacuated and holds it until you are able to go to the loo.

Anus: The last part of the digestive tract, the anus consists of the pelvic floor muscles and two anal sphincters. They provide fine control of the stool and it keeps us from accidentally pooing ourselves; these muscles along with specialised nerves detect if the rectal contents are gas or solid, and help us control the stool until we're able to make it to a bathroom. I say, power to the anus! It is an underrated part of our digestive tract that we should be truly thankful for.

Other important players worth mentioning are:

Enzymes: Proteins or compounds produced by different parts of our digestive system that break down or digest food.

Bile: A funky-looking, yellowish fluid produced by the liver that helps to digest fats in the small intestine.

Pancreas: A funny-looking organ that sits behind the stomach and is an important player in the digestive tract. It produces enzymes that are essential for breaking down food further in the small intestine. You may also know of the pancreas as the producer of insulin and glucagon, two important hormones that regulate our blood glucose (sugar) levels, though these are produced by a different part of the pancreas.

A word (or a few) on movement

We've briefly mentioned how food moves from the oesophagus into the stomach via muscular contractions called peristalsis. It is important to appreciate that there are several programmed muscular movements that are responsible for the transport of food between different parts of our digestive tract.

So, we've started off with peristalsis and now we move on to another type of movement pattern that you should be aware of: the migrating motor complex (MMC). The MMC has been dubbed the 'housekeeper of your gut' and produces a series of cleansing waves to sweep waste from the small intestine. It happens about every 45 minutes to two hours during times of fasting. This movement is deactivated when you eat.

Another wave of movements that occur, this time in the large intestine, is called 'mass movement'. These types of muscular contractions are considered the 'last hurrah', moving waste (faeces) into your rectum ready for defaecation (also known by the popular term 'number twos'). Mass movement occurs several times a day, is on hold overnight and restarts again in the morning, which explains the urge to go first-thing upon waking up.

With this brief tour through your digestive tract, you start to realise how every member, from your mouth to your anus, has a crucial role to play in an act that is sometimes undermined. Your digestive system is a complex network of organs working in perfect synergy, so when one becomes faulty, we may be in a bit of trouble.

How do we develop a faulty gut?

Developing a 'faulty' gut doesn't happen overnight, unless we've picked up a stomach bug causing our souls and insides to exit the building. All jokes aside, what does cause gut issues? If we had to look at the most common causes, there are two important factors, over which we have zero control: genetics and ageing.

Ageing

An unavoidable fact of life, ageing is known to predispose us to gut problems due to age-related changes that occur in different parts of your digestive system. For example:

- Less saliva produced in the mouth and mechanical issues related to chewing can cause difficulties.
- Changes in stomach acidity and elasticity can impact multiple factors, such as how much food we're able to consume and the risk of developing ulcers as well as weakening our stomach's lining, making it more prone to damage.

- Lactose intolerance is a very common occurrence as we age given the reduction in the levels of the enzyme lactase produced in the small intestine, causing an intolerance to dairy products.

- Constipation becomes more common due to a decline in the muscular contractions of your large intestine, causing a delay in the transit of waste.

Then onto genetics...

Genetics

Some gut disorders, such as coeliac disease and inflammatory bowel disease, have a genetic component, meaning they are partially hereditary. By saying that though, being predisposed to the condition is only one part of the equation in most cases. A number of factors have to come into play, such as environmental triggers, and act as the missing switch, leading one to develop the condition.

Factors we can control – the four pillars of gut health

It is not all doom and gloom though – there are a number of factors that we can control that can trigger a faulty gut and they include:

- A diet low in fibre and poor in plant-diversity.
- A diet high in processed and red meat.
- High stress levels that remain unaddressed.
- Being sedentary.

We will be going through each of these factors individually in the course of the book but, by mentioning these points, I just want to plant the seed and get you to reflect on four essential pillars to support better gut health, all of which are under our control to improve. These four pillars are:

1. nutrition

2. the mind

3. movement

4. sleep.

You will be introduced to the four-pillar approach in greater detail when we talk about the specific 'faulty' condition called irritable bowel syndrome (IBS), which is also known as a 'functional' gut disorder.

Now, if you've never come across the term IBS before, it describes a group of gastrointestinal conditions where anatomically, everything looks normal on the outside, meaning, the gut does not exhibit any structural abnormalities that can be seen via x-ray, endoscopy or blood tests. However, the symptoms experienced are the result of an abnormal functioning of your digestive system. In simple terms, there's a glitch in how your gut operates. This group of disorders is classified by a combination of symptoms related to any of the following:

- issues with intestinal motility
- altered immune function
- microbial imbalance (more on the concept of dysbiosis on page 19)
- a 'glitch' in the gut-brain axis.

Functional gut disorders have recently been renamed 'disorders of the gut-brain interaction' and this change has been welcomed as an acknowledgement of the complex interaction of biological, social and psychological factors when it comes to their development. The term 'functional' has left many patients feeling stigmatised since their condition may be viewed as less important or 'real' than that of an organic disease.

How common are gut disorders?

In a global study by the Rome foundation investigating the prevalence of functional gut disorders (FGD) in 33 countries, it was estimated that 40% of adults had to struggle with an FGD. Out of 73,000 people, the survey found that 37% of men and 49% of women met the diagnostic criteria for at least one FGD.

Chances are, if you are reading this book, then you probably have come across someone who lives with a functional gut disorder or it may be safe to say that you suffer from one yourself? Whether you do or just simply want to educate yourself on all things gut, we will be going through different levels of faultiness throughout this book, highlighting the most common woes we may experience, but for now, let's simply appreciate the value and beauty of a well-functioning gut.

On then to the gut's fascinating role of housing a so-called virtual organ of the body, our gut microbiome....

The one about your gut microbiome

At the rate science continues to uncover the wonders of our gut microbiome, I am hoping that this chapter is not going to be too outdated by the time of publication. Yes, our gut microbiome has come a long way this past decade and more and more discoveries are likely.

The world of our gut microbiome

If you're new to the term, the 'gut microbiome' refers to the tiny microbes, mostly bacteria but also viruses, fungi and yeasts, that live in our digestive tract, including their genetic material. When we refer to them as 'gut microbiota', we are referring to them as a community of microbes living together (excluding their genetic make-up). In general parlance the two terms are used interchangeably as the distinction is useful mainly for research. In this book I will take about the 'microbiome' to be as all-encompassing as possible.

It is estimated that about 100 trillion microbes exist in this complex and beautiful environment in each one of us, which researchers have now dubbed a virtual organ of the body. This means that there are more bacterial cells than human cells in your body so you can think of us humans as big walking blobs of bacteria.

Now, we have known for a long time that we had microbes living inside of us but back in the day, and I'm talking way back, we had no idea what they were, what they did or how many of them there were. Thanks to technology and evolving science, we are now able to study the DNA of these microbes, find out that we have over 1000 different types of bacteria living within us and that they are more important than we thought when it comes to our overall wellbeing and the health of every organ within us.

One of the big questions that scientists have debated has been, where does our gut microbiome come from? For years, we'd always believed that the womb was a sterile bubble and that exposure to our first microbes only happened as we emerged into the world via our mother's birth canal. Some recent studies have now shown that babies may actually be exposed to microbes before they are even born, so our first dose of gut bugs potentially happens way before our grand entrance.

Now, babies born vaginally are covered in a microbial coating as they travel through the birth canal, and these first microbes appear to have a huge role in the development of our immune system.

Babies born via C-section are said to be exposed to their first proper set of microbes from their mother's skin and the outer environment that they find themselves in. There's also been a lot of talk about whether babies born via C-section are more prone to asthma and allergies because of the types of microbe they've been exposed to at birth, but luckily, breastfeeding seems to reduce that risk.

As we grow, our mother's microbes aren't the only ones influencing our own gut microbiome; breastfeeding, diet, our environment, exposure to pets, stress, medication... and much more all end up shaping our inner ecosystem.

So, who are the big guys when it comes to our gut microbiome? Bacteria are actually the most predominant inhabitants of our digestive tract, mostly colonising our large intestine. Let us for a moment, appreciate the many things they do. They:

- digest food components.
- absorb and produce nutrients such as B vitamins, vitamin K and health-producing compounds called short-chain fatty acids. (A whole section has been devoted to these in Chapter 4 - see page 40.)
- produce hormones and neurotransmitters - that is, chemical messengers that send signals to the brain.
- maintain the protective barrier that lines our digestive tract.
- provide immunity.

The last decade has thus uncovered quite a lot. The most researched areas of our gut microbiome's roles include:

- How it impacts our whole biology and different body organs via the 'gut-brain axis' (see page 20) - I want you to think of this bidirectional communication pathway as if your brain and gut had access to an 'unlimited call package' chatting away, all day, constantly, about how everyone else is doing.

- How its role as guardian of our inner galaxy functions – how it somehow controls our immune system by communicating with immune cells on how to defend us and respond to infection. We now know that 70% of our immune system lies in our gut.

- How it contributes to health and disease, including to diabetes, obesity, inflammatory bowel disease (IBD), autoimmune disorders, allergies and neurological problems.

- How it influences our mental health, an area I am truly passionate about, where studies are now looking at links between the health of our gut microbiome and depression, anxiety and Alzheimer's.

Drawing on this research, I want to take an in-depth look at the different communication pathways between the gut microbiome and other parts of the body as it's essential to highlight the connections we are slowly uncovering, thanks to science, and how different systems in our body are being influenced by the oh-so magnificent gut microbiome.

Let's kick things off with the gut-immunity axis and then look at the gut-brain axis, gut-thyroid axis, gut-skin axis and gut-lung axis, each showing how fundamental gut health is to every aspect of our overall health.

The gut-immunity axis

There's so much talk about how we can 'boost our immune system', and no more so than during the COVID-19 pandemic when I started writing this book. However, it's important to know that, firstly, a 'boosted' immune system is actually not a good thing as that implies it has gone into overdrive and is overreacting. This has become such a huge marketing term that it makes me cringe when I hear lots of 'super-foods' being promoted as essentials for immunity

– enter garlic, turmeric and some magical leprechaun concoction that hasn't hit the shelves yet. When it comes to immunity, there are essential pillars that support it: nutrition, sleep, stress management and movement. These pillars also influence the health of our gut microbiome and our overall wellbeing so we need somehow to change our approach when it comes to immunity and think of our gut rather than of single 'super foods.'

If we had to look at our gut microbiome's role specifically in relation to immunity, we can formulate a list of important observations:

1. Our immune system is regulated by our gut bacteria from birth.

2. Achieving a healthy bacterial balance in our gut is essential for a well-functioning immune system.

3. An imbalanced gut microbiome – e.g. having more harmful than beneficial bacteria – can lead to disease and cause an inflammatory state. Enter the concept of **'dysbiosis'**. Just as with everything in life, balance in your gut is the key to success. Dysbiosis can be seen as an unfavourable state that may involve one or more of the following:
   • Reduction in microbial diversity.
   • Changes in the types of beneficial species.
   • Changes in how beneficial microbes function.

Dysbiosis has played a possible part in a variety of health conditions, including, inflammatory bowel disease (IBD), irritable bowel syndrome (IBS), type 2 diabetes, rheumatoid arthritis and mental health disorders. Now oftentimes, dysbiosis can be prevented. Some of the causes of dysbiosis include:
   • Antibiotic use.
   • An unbalanced diet lacking in nutrients and fibre or containing harmful substances.
   • Alcohol abuse.

- High stress levels.

4. Our immune system would not be fully functional without the power of GALT, or 'gut-associated lymphoid tissue'. This is located within our digestive tract and is the reason why we consider 70% of our immune system to be located in our gut. GALT is responsible for identifying who is 'friend or foe' as well as regulating inflammatory and allergic responses.

In the next chapter, I'll be going through ways for supporting this axis through nutrition and lifestyle, so keep flipping those pages.

The gut-brain axis

Another important feature of our gut microbiome is how it communicates with our brain. Your brain and your gut are physically connected by this long tubular structure called the vagus nerve. I now want you to think of your brain and your gut chattering away, constantly communicating via this channel. One way in which these chats take place is via chemicals (neurotransmitters), produced by gut bacteria, that travel through the vagus nerve directly to the brain, influencing your mood. Through the same channel, your brain can also send signals directly down to your gut.

Here are a few more fun facts that can back up the claim that 'your gut is your second brain':

- The brain and the gut are the only two organs with their own nervous systems. In your gut, it is called the 'enteric nervous system'.
- There are more neurotransmitters in your gut than your brain.
- Altogether 50% of the dopamine (aka the calming or 'feel good' hormone) in our bodies and 90% of the serotonin (aka the happy hormone) is produced in our gut.

These ground-breaking discoveries, highlighting how our gut microbiome is involved in more than just 'gut stuff', are showing us that our gut microbes can shape and influence brain development and behaviour and may contribute to mental illness. Also, certain emotions, feelings and 'states' of being can influence our gut – for example, stress and anxiety can cause bloating and diarrhoea in many of us. Stress can have a negative impact on our gut in many ways, as mentioned:

- It can affect how our gut muscles contract.
- It can cause a glitch in the gut-brain axis.
- It can negatively impact the health and survival of our gut microbiota.
- It can reduce the healing capacity of our gut.

This serves as a little nudge to all my readers never to neglect your mind when it comes to addressing gut issues. Chapter 12 has you covered; it will take a deep dive into this mind-gut connection.

The gut-thyroid axis

Given how many clients who have consulted at my clinic suffer from a thyroid disorder called hypothyroidism, and how widespread thyroid disorders are worldwide, it is not surprising that a great deal of research is currently underway trying to better understand the gut-thyroid connection. Hypothyroidism is the term for an underactive thyroid and this is thought to affect around 200 million people around the world.

In case basic biology is not your forte, the thyroid is a butterfly-shaped gland found in your neck that is part of your endocrine (hormone) system. It produces thyroid hormones that are responsible for essential bodily processes such as respiration, all aspects of your metabolism and regulating your reproductive function and hormonal health.

Hypothyroidism is the commonest thyroid problem, and its most frequent cause is an autoimmune disease called Hashimoto's thyroiditis. As with all autoimmune disorders, your immune system becomes triggered to produce antibodies that attack your own tissues. In the case of your thyroid, the immune system attacks your thyroid gland impacting its ability to produce thyroid hormones and thereby disturbing its function.

As with every autoimmune disease, you'll find lots of talk around 'heal your gut, cure your autoimmune disease' but this is an overly simplistic message. By saying that though, yes, the gut should not be neglected when it comes to managing such conditions and here's why:

- Just in case it hasn't yet sunk in, 70% of your immune system lies in your gut.

- When your immune defences are compromised, that can increase your risk of inflammatory conditions, in turn, causing potential issues with your gut lining.

- The exact causes of why that happens are varied.

So where does the idea that the gut-thyroid connection is important come from? We do know that thyroid and gut conditions can coexist – such as in the case of Hashimoto's and coeliac disease, both autoimmune conditions; anyone diagnosed with Hashimoto's would benefit from being screened for coeliac as there is definitely an increased risk. We also see links between our gut microbes and the roles they play in the production or functioning of our thyroid hormones. We also know that people with autoimmune conditions show poor microbial diversity, meaning their inner ecosystem is not as varied in terms of species, or may lack some important anti-inflammatory species, or 'soldiers'.

The takeaway message? Gut health may indeed be central to our wellbeing, especially when it comes to dealing with autoimmune disorders.

The gut-skin axis

We cannot ignore the connection between our gut and the biggest organ of our body, our skin. Our skin is the first protective barrier against physical, chemical and bacterial hazards in our environment. Just like our gut, it houses millions of microbes that look after our skin's health. When it comes to the development of skin disorders, dysbiosis can take place within our skin microbiome but we are now seeing how certain skin conditions are also accompanied by an imbalanced gut microbiome. This connection, dubbed the gut-skin axis, has been known to us since as early as the 1930s.

Animal and human studies have started to tease out this connection by looking at how the gut microbiome can contribute to common skin disorders such as acne, atopic dermatitis and psoriasis. Certain gut microbes and their metabolites (i.e. compounds they produce) can generate an anti-inflammatory or pro-inflammatory response, promoting the development of these skin conditions or positively aiding their management. This certainly demonstrates that, when targeting skin conditions, we should shift our focus from purely relying on topical ointments to understanding how to nourish our inner ecosystem to produce more anti-inflammatory, skin-healing metabolites.

The gut-lung axis

As much as we no longer want to hear or read about the pandemic that changed the world, mentioning the big C in this book was inevitable. Covid-19 has opened up a portal into the world of the gut-lung axis, which is the bidirectional communication pathway between our lungs and our gut. The cross-talk between the two had been investigated as early as the 70s, so we're not really marching on ground-breaking territory, but there are still so many unknowns.

Just like our gut and skin, our lungs also have their own unique ecosystem, but it is not as dense and diverse as our gut's microbiome. These microbes are responsible for the cross-talk that occurs between the organs via chemical messengers they produce, ultimately affecting immune responses in the lungs and gut. As we've seen with other respiratory disorders such as influenza and COVID-19, gastrointestinal symptoms (such as nausea, diarrhoea and vomiting) can occur together. An interesting and significant observation that studies have reported is that patients with COVID-19 have reduced levels of beneficial gut microbes while those of a pathogenic nature are high.

In the course of this book you will get to know some of these microbes on a first-name basis, but for now, I want you to think about how looking after your gut is indeed key to your overall well-being given just the few connections we've gone through in this chapter. Or, perhaps I am biased trying to convince you that this is the case since I've been working in this field for most of my career!

With all this dysbiosis talk, you may be tempted to get your hands on a microbiome analysis kit to get an idea of how balanced your inner ecosystem is or whether you're missing out on some beneficial microbes. Before you invest in one of these testing kits, I want you to know that there actually is no 'ideal gut microbiome' to strive for. Think of how unique our fingerprints are and you can look at your gut microbiome similarly.

But Sandra, what is a 'good looking' microbiome you may ask? The single most important feature of the microbiome is DIVERSITY. A diverse gut microbiome (meaning having a wide variety of different types of microbe in differing amounts) is what we should strive for. The big question here is: 'Does a low diversity of microbes make us sick?' What we are now seeing is that many conditions, such as diabetes, coeliac disease, allergies, obesity, inflammatory conditions, depression and anxiety (that's only a short list!), may be associated with low microbial diversity.

In the next chapter, we'll be going through strategies for boosting microbial diversity by understanding how to feed our gut microbiome. But first, and as promised, shall we meet some of these tiny critters that reside within us?

Introducing the key players

Bacteria are scientifically classified into primary groups – phyla – and these are then broken down further into the subgroups of genus, species and strain. The two phyla Firmicutes and Bacteroidetes represent about 90% of our gut microbiota. When we move down the classification hierarchy to genus, which is basically the 'first name' of a bacterium, we have Lactobacillus or Bifidobacterium, for example. Species is the 'second name' – e.g. acidophilus – and strain is even more specific and is usually noted as a number helping us distinguish between microbes that are part of the same species – LA5, for example.

In a recent landmark study called PREDICT, researchers were able to identify 15 species of gut microbe that were linked to positive health markers and healthier diets and 15 microbes that were associated with poor health and unbalanced nutrition. We are nowhere near revealing and naming every one that resides within our digestive tract as there are so many more gut microbes we've yet to discover. For now, let's make the acquaintance of some popular microbes, starting with Lactobacillus.

Gut health glossary

As you read through, you'll come across some frequent terminology so to clear up any confusion:

- **Probiotics:** Live microorganisms that, when consumed in adequate amounts, confer a health benefit on the host (us!).
- **Prebiotics:** Types of fibre that are food for the good bacteria (i.e. probiotics).
- **Postbiotics:** Beneficial bioactive compounds produced by our gut microbes during the fermentation process.

Lactobacillus (L.)

You might've come across these microbes as probiotic supplements in the form of capsules or sachets, or on live-yoghurt ingredients. Lactobacillus is a genus of bacteria, containing about 300 species, of which there are countless strains that have been studied. These microbes are essential members of our gut microbiome as they have been shown to either prevent or manage different conditions that can affect our digestive system. Numerous Lactobacillus species and strains have been shown to benefit gut conditions such as irritable bowel syndrome (IBS), ulcerative colitis, watery diarrhoea, bloating and allergies as well as mental health conditions such as depression and anxiety. They include:

- *L. rhamnosus* GG
- *L. plantarum*
- *L. casei*
- *L. helveticus* R0052
- *L. acidophilus* NCFM.

I'll be mentioning some of these strains throughout the book when we talk about the use of probiotics in managing specific gut conditions.

Bifidobacterium (B.)

A household name in the gut microbes' department, Bifidobacteria are among the first microbes to colonise our gut and are crucial inhabitants that stabilise the gut microbiome. They are famously known for producing a number of powerful compounds, such as:

- Vitamin K and several B-vitamins.
- Short-chain fatty acids (SCFAs) aka postbiotics.
- Antimicrobial compounds that act as artillery against pathogens.

Bifidobacteria species and strains have been shown to protect against inflammation and other conditions such as colorectal cancer, diarrhoea and constipation, inflammatory bowel diseases (Crohn's, ulcerative colitis) as well as eczema. They are also well known for regulating our immune system and are being investigated for promoting mental wellbeing. Some of the big contenders that have been extensively researched include *B. lactis, B. longum, B. animalis* and *B. breve*.

Faecalibacterium prausnitzii (F. prausnitzii)

An absolute mouthful to pronounce and remember, *Faecalibacterium prausnitzii* is one of the most abundant bacterial species in healthy guts. One of its main roles is the production of the SCFA (aka postbiotic) butyrate, which fuels the cells along our intestinal lining, prevents inflammation and can also trigger the production of hormones that regulate our appetite. Butyrate is a postbiotic that will be mentioned A LOT in this book. Low levels of *F. prausnitzii* have been linked to conditions such as IBD, diabetes and chronic fatigue syndrome (CFS).

Clostridium (C.) species

And enter the 'bad' guys... Before we talk about pathogenic gut bugs, it is important to note that 'bad' microbes are indeed present within our inner-ecosystem but, under normal circumstances and when our microbiome is not in dysbiosis, these microbes rarely cause any harm. However, when the opportunity arises, allowing them to over-populate and crowd out the beneficial microbes in our gut, we have a major 'Houston, we have a problem' situation.

Clostridia are known to produce the largest number of toxins of any bacterial class and I am certain you've come across *Clostridium difficile* or *C. difficile*, which is responsible for causing horrible diarrhoea and colitis (inflammation of your colon) and can have serious consequences for those with a weak immune system and the elderly. *C. difficile* infection is linked to antibiotic use and hospital stays but keep in mind that antibiotic therapy doesn't immediately put you at risk of infection. Antibiotics are meant to eradicate problematic microbes, but they end up wiping out the good guys too. What are the risk factors for antibiotics allowing *C. difficile* to take over?

- Long-term use of antibiotics.
- Being over the age of 65.
- Having a suppressed immune system.
- Long-term use of anti-acid medications called proton pump inhibitors, such as Omeprazole.

Treating *C. difficile* involves using the exact medication that causes it – antibiotics. However, using faecal transplants or FMT (yes, it's totally a thing! – see the box on page 29) is proving to be more effective given that reinfection is common and that these microbes tend to develop resistance to antibiotics.

A note on FMT – Faecal microbiota transplantation

In simple terms, FMT is an emerging treatment involving a poo transplant with the aim of repopulating the gut with healthy microbes. The theory behind it is that the procedure – a poo transplant from a healthy pre-screened donor into the sick patient – addresses dysbiosis and the poor microbial diversity that may be responsible for the illness experienced. Currently, treatment is approved only for the management of C. difficile infection, but its success rate has prompted interest in how it can be used for other conditions such as IBD, IBS and metabolic syndrome. A word of warning though, NEVER ATTEMPT FMT AT HOME!

Candida albicans (C. albicans)

Remember how we spoke about our gut microbiota's inhabitants being mostly bacteria but also fungi? Meet the 'yeast' of them all, *C. albicans*. It is the most prevalent fungus in our digestive tract and, in healthy bodies, causes no problems. It is part of our oral, vaginal and gut microbiota and its overgrowth, or 'candidiasis', can result from antibiotic use where the good bacteria that keep candida in check are killed off. Furthermore, those with a suppressed immune system can be at higher risk of developing a yeast infection.

Candidiasis of the mouth and throat is also known as thrush and symptoms include the formation of a white coating inside your mouth (tongue and inner cheeks) together with soreness and redness. Saying that though, the most common location for candidiasis is actually the vagina, aka vaginal thrush, and symptoms include unusual discharge and itching. Despite what unicorn and wellness gurus may preach, candida overgrowth in your gut is not as common as they would like you to believe and, no, symptoms such as headaches, brain fog and chronic fatigue are not symptoms of candida overgrowth in the gut.

C. albicans can be detected in poo tests of healthy individuals and, as I've mentioned, that's not necessarily a problem; the last thing you should be doing is jumping on an anti-Candida diet or cleanse. If yeast overgrowth in your gut is problematic, common symptoms include stomach pain, bloating, nausea and irregular bowel movements, but these can overlap with a ton of other gut issues and your doctor will be able to guide you towards the right diagnosis (and by doctor, I mean a GP or gastroenterologist, NOT Google nor Dr Celebrity MD on Instagram trying to sell you an anti-Candida detox).

4

The one on feeding your gut microbiome

It's lunch time and Jemma picks up her usual meal choice and heads back to her desk where she multitasks, feeding herself, answering emails and investing in a thumb workout, scrolling through different social media accounts promising the holy grail of gut health.

Not realising that she's impregnated herself with so much pseudoscience and misinformation, she is totally hooked on the fact that she needs coffee enemas, a side of celery juice, and apple cider vinegar gummies to 'heal her gut' (and oh, don't forget the sage).

Jemma's story, my friends, is the perfect example of how social media has become the new healthcare and when it comes to nutrition, social media is transforming it into a religion.

How we nourish our bodies has a huge impact on the health of our gut microbiome – certain foods may help specific gut bugs to thrive, but the way we eat and live may also cause many to perish, creating an unfavourable world, possibly leading to disease. An unbalanced gut microbiome (aka dysbiosis) has been linked to around 80 conditions, but we're still only scratching the surface, not fully understanding the link.

However, here's what we do know:

- A diverse gut microbiome (meaning having a wide variety of different types of microbe in differing amounts) is what we should strive for – a good way to think about this is to treat your gut microbiome as a garden that needs soil, fertiliser and care to thrive and this is where food and lifestyle come into play.

- A diverse diet, predominantly plant-based and high in fibre, is the ideal basis for achieving a thriving garden.

- We also know that there are other pillars of gut health that we should not neglect and these are mind, movement and sleep, all of which can directly or indirectly affect our garden. For example, have you ever experienced the urge to run to the bathroom when you're stressed or anxious? Or had 'butterflies' in your stomach? That's basically the simplest example of how the mind and gut are connected and how addressing the mind should not be overlooked.

If I haven't lost you already, I want us to dive head first into the nutritional aspects of developing a thriving garden. But before looking at how to feed your gut microbiome, we need to go through some general nutrition 101s.

Introduction to the building blocks of food

Most foods can be classified according to the three major (or 'macro') nutrients they provide: carbohydrate, protein and fat. Each of these affects our bodies differently. Kicking things off with the currently most misunderstood, demonised nutrient of them all, let's talk about carbohydrates.

Carbohydrates

We have recently called them 'bad', 'fattening', 'the devil' and so on, but the true nutritional benefit of carbohydrates has been jeopardised by popular myths and misinformation. Yes, you can eat carbohydrates after 5 pm and, no, they are not fattening.

Carbohydrates are food compounds that can be classified into three types: starches, sugars and fibre. They are also known as the body's energy foods as they provide fuel for all the body's activities. Carbohydrate foods are broken down into glucose (a type of sugar) and other simple sugars such as fructose and galactose. This glucose is absorbed into the bloodstream and acts as a fuel for our body, and the pancreas secretes a hormone called insulin to help glucose move from the bloodstream into our cells.

Carbohydrates are the main driver of our blood sugar levels. Where do we find them?

Breads	Milk
Breakfast cereals	Noodles
Flours	Pasta
Fruit	Rice
Fruit juice	Starchy vegetables: potato,
Grains	sweet potato, corn
Honey	Table sugar
Legumes (e.g. beans,	Yoghurt
chickpeas, lentils)	

Let's clarify one thing: carbohydrates DO NOT CAUSE WEIGHT GAIN. You gain weight when you consume too many calories, whether they are from carbohydrates, fats or proteins. Low-carb diets lead to SHORT-TERM weight loss because:

- Weight loss is mainly water and not body fat.
- Calories are restricted.

I've never been a fan of low-carb diets because they tend to deprive people of my favourite F word, fibre. We're going to talk about fibre in A LOT of detail so keep reading.

When it comes to carbohydrates, consider the following questions:

How much?

What type?

What goes on it/do you eat with it?

The amount of carbohydrate that you should have per day usually depends on your activity levels and that is how I usually set my clients' daily allowance of carbs.

Fat

Fats are another macro constituent of food, also known as lipids. They form a major part of all cell membranes in our body and are essential for the absorption of the fat-soluble vitamins A, D, E and K from foods. There are different kinds of fats in our diet, found in foods such as avocado, butter and margarine, animal meat, milk and milk products, nuts and seeds, and oils.

As we have different types of fat in our diet, some better than others and some essential, let's talk about the fats that may negatively influence our gut microbes a few pages down (see page 37). Before that, the three main types of fat we need to be aware of are: saturated, polyunsaturated and monounsaturated.

Saturated fat

Saturated fat is usually solid at room temperature and is found in animal foods like fatty meat, milk, butter, cream and cheese as well as coconut and palm fat/oil. A small amount of saturated fats in our diet won't do any harm but we should aim for no more than 10% of our daily calories coming from this type of fat. At the same time, we should be focusing on getting the right balance of polys and monos, which are described below. There has been a lot of controversy surrounding saturated fats and whether they are responsible for raising our LDL ('bad cholesterol') levels, in turn, increasing the risk of heart disease and stroke. By looking at the evidence, we still cannot promote saturated fats as innocent bystanders when it comes to increasing our risk of heart disease and blood clots. One of the largest studies to date has shown the positive impact of reducing these types of fat on heart health but my takeaway message is this: remember that foods work in synergy and we should look at our overall dietary pattern rather than obsess about one specific food component.

Sources of saturated fats include:

- Full-cream dairy such as full-fat milk, hard cheese, butter, ghee and full-fat yoghurts.
- Fatty cuts of lamb (e.g. chops), fatty beef and pork (bacon, crackling).
- Chicken cooked with the skin on, chicken skin and fat, duck.
- Processed meat products such as sausages and salami.
- Dripping, lard, copha.
- Pastries, croissants, pies, commercial biscuits and cakes.
- Chocolate, cheesecakes, donuts.
- Potato crisps.
- Fried take-away foods.
- Palm oil, coconut oil, coconut cream, coconut flesh.

Polyunsaturated fats

Polyunsaturated fats exist in two forms – omega-3 and omega-6 fats. These beneficial fats are also known as 'essential fatty acids' as our bodies are unable to produce them. For this reason, we must include them as part of a balanced diet. Aim to include moderate amounts of foods rich in polys such as:

- Fish – mackerel, tuna, salmon, trout, sardines, trevally, pilchards.
- Sesame and soyabean oils, wheat-germ oil, linseed (flaxseed) oil.
- Walnuts, Brazil nuts, sunflower seeds and sesame seeds.
- Soya milk, soya yoghurt.

Monounsaturated fats

Monounsaturated fats can help lower blood cholesterol levels while maintaining 'good' HDL cholesterol levels. Foods high in these fats include:

- Extra-virgin olive oil

- Avocado

- Peanuts

- Some nuts: macadamias, pistachios, almonds, hazelnuts, pecans.

Note: Seeds, nuts, nut spreads and peanut oil contain a combination of polys and monos. However, such foods should be consumed in moderation as they have a high calorie content.

What about trans fats?

Trans fats are formed by heating liquid vegetable oils with hydrogen in order to make them solid, a process called hydrogenation. They are found in margarines, processed foods, snack foods and commercially-fried foods. Just like saturated fats, trans fats raise your 'bad cholesterol' levels (LDL) and also lower your 'good' HDL cholesterol levels. Nowadays, food manufacturers have adopted new production methods to reduce levels of trans fats.

Most countries have banned foods containing more than 2% trans fats, while others require the trans-fat content to be included on food labels.

A word on cholesterol

Cholesterol is a fatty wax-like substance that is produced in the body, in the liver. The term 'cholesterol' may hold negative connotations but, folks, cholesterol has some pretty crucial roles when it comes to the functioning of our biology. It helps form the structure of our cell membranes, is involved in the production of hormones and is needed to create bile acids, which help us digest dietary fat. Keeping things simple, there are two types of cholesterol as you may be aware – HDL cholesterol ('good') and LDL cholesterol ('bad').

- HDL, aka high-density lipoprotein, is known as the good cholesterol because it carries cholesterol from different parts of your body, back to your liver where it can be removed.
- LDL, aka low density lipoprotein, transports cholesterol from the liver to areas in your body where it may be needed, such as in the production of sex hormones. It has been dubbed 'bad' because high levels of LDL circulating in your bloodstream can lead to a build-up of cholesterol (plaque) in the blood vessels.

What you need to know is this – when you eat foods containing saturated or trans fats, your body will produce the different types of cholesterol. If you consume a lot of these fats, your body will produce a lot of LDL cholesterol. This could increase your risk of heart disease and stroke if the build-up of plaque starts to occur. Luckily, we can always get our blood cholesterol levels checked to see where we currently stand. Solution? Be mindful of how much saturated fat you are consuming and include more polys and monos.

Protein

Protein is another fundamental nutrient. The building blocks of proteins are called amino acids. Proteins are used to build each body cell, such as those that make up our muscles, bones, hair, blood and much more. They are important in cell growth, repair and maintenance, making them essential throughout life. In total, we have 20 amino acids that join together, in different combinations, to form proteins. Out of these 20, nine are considered 'essential' amino acids, meaning we need to obtain them through food as they cannot be made by our bodies. The other 11 are called 'non-essential' since our bodies are able to produce them.

Protein-containing foods include the following, noting that some also contain carbohydrate, as indicated with '*':

Beef	Milk*
Cheese	Nuts
Chicken	Pork
Eggs	Seaweed, microalgae
Fish	Soya milk*
Grains (e.g. quinoa, brown rice)	Tofu*
Lamb	Turkey
Legumes* (beans, lentils, chickpeas)	Yoghurt*

Understanding the basics of a wholesome diet is essential to help you construct and maintain well-balanced meals throughout the day. The term 'balance' implies that you include a variety of foods from the major food groups and they include:

- Breads and cereals
- Meat and meat alternatives
- Fats
- Fruit
- Vegetables
- Dairy products and their alternatives.

So, here's a quick recap. We've got important food components to include daily – carbs, proteins and fats – and when we mention the term balance, we are talking about including all the major food groups daily in amounts that are suited to an 'individual's' needs.

Now, let's move on to the final two food components I want to describe: fibre and polyphenols.

Fibre, our favourite 'F' word

Whenever the word fibre pops up, an oh-so popular emoji pops into everyone's mind. Yes, faeces, poo, stool is always associated with the word fibre, but what actually is FIBRE?

Fibre is the backbone of all plant foods and it passes through our digestive system, relatively unchanged until it reaches our large intestine. Where is it found? Think fruit, vegetables, nuts, seeds and wholegrains. Fibre is more important to our gut microbes than it is to us but it does offer our species some huge benefits:

- It adds bulk to our stools, making us poo regularly and with ease.

- It binds to certain compounds, regulating our blood sugar levels better and lowering cholesterol.

- It reduces the risk of certain diseases including colon cancer, heart disease and type 2 diabetes.

Once fibre reaches our gut microbes in the large intestine, it is fermented into these health-promoting by-products called short-chain fatty acids (SCFAs). The three main types of SCFA are acetate, butyrate and proprionate. Why so important you may ask? SCFAs offer first-class comfort when it comes to poo, meaning they improve laxation, reduce the risk of colon cancer and fuel the cells that make up the intestinal lining. By being the golden fuel for these cells, SCFAs make them better guardians when it comes to protecting us against the nasties.

Types of fibre

In earlier times, fibre was broken down into three main kinds:

Soluble fibre: Slows down digestion and absorption. It helps improve blood cholesterol and sugar levels while having prebiotic properties (see page 288). Food sources include beans, lentils, chickpeas, soya beans, oats, barley and fresh fruit and vegetables.

Insoluble fibre: Adds bulk to your stools, making them softer and easier to pass. This type of fibre is great for preventing constipation and bowel conditions. You can find insoluble fibre in the skins of fruit and vegetables, wheat bran, wholegrain breads and cereals, brown rice, nuts and seeds, and flaxseed.

Resistant starch: Has a similar function to soluble fibre in helping slow down digestion. It is golden fuel for your gut bacteria, with prebiotic properties. This type of starch can be found in unripe bananas, legumes, raw oats and cooked and cooled pasta, rice, potatoes, cashews, sorghum and millet.

You'll have noticed a recurring word here – PREbiotic. Hoping I won't lose you when I get into this, but prebiotics are fibres, though not all fibres are prebiotics.

Having 'prebiotic' properties is like being endowed with a superpower. For a fibre to be considered 'prebiotic', studies need to have shown that it can actually influence the health of our gut microbes. As a reminder of the terminology (see box on page 26), think of prebiotics as the food for probiotics; as such, they influence the balance of our gut microflora that ferment them into SCFAs. And probiotics are 'good bacteria' that are similar to the ones found in our gut and offer health benefits by repopulating our intestinal microflora and thereby positively influence overall balance.

If we consume both probiotics and prebiotics at the same time, commonly in 'supplement' form, we have a dynamic duo dubbed 'synbiotics'.

Now, remember how I initially said we used to divide fibre into three main types? Recently, more and more researchers have been suggesting we divide fibre into categories based on physical characteristics:

- Solubility (how well it dissolves)
- Viscosity (how it thickens)
- Fermentability (the extent to which it is broken down by our gut bacteria).

This brings us back to prebiotics; the three main prebiotic fibres are:

1. fructo-oligosaccharides (FOS)

2. inulin

3. galacto-oligosaccharides (GOS).

We also have pectin (a type of soluble fibre found in fruit) and arabynoxylan (the main soluble fibre component in cereal grains).

Why should we shift our focus from probiotics to prebiotics and postbiotics?

Prebiotic fibres help us to maintain an ideal, happy balance of gut microbes and to promote the production of SCFAs. The most popular SCFA, which has gained a ton of attention, is butyrate, as mentioned earlier. If we had to sum up butyrate's benefits, they include:

- Fuelling the cells of your colon (called colonocytes). Without butyrate, these colonocytes, that make up your gut lining would not function properly.

- Suppressing inflammation. Studies have shown that butyrate possesses anti-inflammatory and anticancer properties, where it can 'neutralise' or zap pro-inflammatory compounds, stripping them of any weaponry that can cause us harm.
- Managing blood sugar levels and maintaining a healthy body weight by influencing the production of gut hormones that play a part in improving blood sugar levels.
- Influencing brain health and protecting our nervous system.

Given these roles, no wonder that certain diseases are now being associated with low butyrate levels. These conditions are insulin resistance, inflammatory bowel disease (IBD) and bowel cancer. The bottom line is, we need more butyrate-producing microbes and to help them thrive by feeding them a diverse mix of prebiotics. Table 1 gives a more useful way to remember these prebiotics fibres, by looking at their food sources.

Table 1: Sources of different prebiotic fibres

FOS: Fructo-oligosaccharides	GOS: Galacto-oligosaccharides	Inulin	Pectin	Resistant starch
Asparagus	Black beans	Artichokes (globe, Jerusalem) Asparagus	Apples	Barley
Beetroot	Borlotti beans	Banana (green)	Banana	Cashews
Bran (wheat)	Butter beans	Barley	Beetroot	Cassava root
Bread (wholemeal, pumpernickel)	Butternut pumpkin	Chicory root	Black-berries	Chickpeas

Table 1 (cont'd)

FOS: Fructo-oligosaccharides	GOS: Galacto-oligosaccharides	Inulin	Pectin	Resistant starch
Cashews	Chickpeas	Garlic	Blue-berries	Cooked and cooled pasta, potatoes, rice
Fennel bulb	Green peas	Leeks	Carrots (raw)	Plantains
Garlic	Lentils	Onion	Eggplant	Red kidney beans
Onions	Mung beans	Cooked and cooled pasta	Grapefruit	Taro root
Pistachios	Pinto beans		Green beans	White beans
Snow peas	Pistachios		Kiwi fruit	
Watermelon	Soya beans		Sweet potatoes	
			Zucchini/ courgette	

How does this translate into the gut microbiota world? Here's an example:

Faecalibacterium prausnitzii was one of the good gut bugs introduced earlier (page 27) and, given that it is a butyrate-producing microbe, we want lots of it so we can get more butyrate! To do that we need to nourish it with a mix of prebiotic fibre such as inulin, pectin and FOS. That means its preferred menu would include pistachios, pumpernickel bread, barley, asparagus, green peas and sweet potatoes.

Do you need a probiotic supplement?

My 'food first' policy encourages all my clients to shift their focus from probiotics to prebiotics (now you know why), but I still get asked whether people should include a probiotic supplement for good gut heath.

I hate to break it to you, but not everyone will benefit from probiotic supplementation. Probiotic supplements have been heavily marketed as a must-have or a cure for all ailments. If you have a good gut, there is no need to be taking these. So, who would actually benefit? Research shows that probiotics may help with:

- Constipation
- Diarrhoea from antibiotic use
- Diarrhoea from an infection
- Irritable bowel syndrome (IBS)
- Ulcerative colitis.

One of the most important things to consider with probiotic supplementation is the need to treat them like medication, meaning that you need to know the type of probiotic strain that is required to treat your health condition. It is the strain of the bacterium or yeast that is most important and determines if the product is going to work. Probiotic supplements should be taken for anywhere from four to 12 weeks at a time and it is also recommended to take about 10 g of prebiotics per day at the same time. For example:

- If you suffer from constipation then an appropriate strain would be *Bifidobacterium lactis* DN-173010 or *Bifidobacterium lactis* BB-12.
- To prevent diarrhoea from antibiotics, choose *Lactobacillus casei* DN-114001 OR *Saccharomyces boulardii*.

Please note, if you are suffering from a weak immune system or major illness, are pregnant or an infant, you should speak to a health professional before ingesting probiotics. Furthermore, be careful if you are allergic to soya or milk. Some of the supplements are grown using strains of soya and/or milk protein.

Sandra's suggestions

- Ultimately, do not take probiotics unless you have discussed this with a health professional.
- Make sure that you read the label to check if a specific probiotic is suitable for your diagnosis.
- Take at least 500 million CFU/day to benefit from a probiotic supplement. However, the dose can vary for different health conditions.
- Make sure that the product you are buying is certified and has been researched, as there are a variety of providers out there just looking for your money.
- Do not take the supplements for more than 12 weeks unless you've been told otherwise by your doctor, gastroenterologist, clinical dietitian or nutritionist.
- And last but not least, make sure that you have a varied and nutritious diet with colourful meals, high in fibre, wholegrains, fruit, vegetables and legumes. This will also help to create greater diversity in your gut flora.

If you're generally healthy with no gut issues, forget about wasting your money on supplements and get some natural sources of probiotics into your system. You can then start adding more prebiotics into your diet too. (Remember the dynamic duo of prebiotics and probiotics.)

Fermented foods as probiotics

Fermented foods can be a natural source of probiotics, but for a product to be classified as a probiotic, the cultures in it should be in numbers sufficient to offer a benefit to the host – meaning us. The process of fermentation involves the use of good bacteria, such as Lactobacillus (page 26) and Bifidobacterium (page 27). When it comes to fermented foods, cheese and yoghurt are the simplest examples; they have been around for many years and are usually the easiest way for consumers to add probiotics to their diet.

The world of fermented foods is complicated because there are no conclusive human studies to show they are an essential part of our diet, but from what we know, they're not bad either. I would still suggest you include a small amount of fermented foods daily only if they don't cause you problems as many people with IBS may find them difficult to tolerate. I've listed some options below; they include things like sauerkraut and kimchi, which are types of fermented cabbage; kefir, which is milk fermented with kefir grains; and miso, which is a paste made from fermented soybeans, but yoghurt is an example of an ancient fermented food. When I'm talking about small amounts, I mean about a tablespoon of sauerkraut or 50 ml of kefir daily, or even every other day. Remember, every person is different in their preferences and in what they can tolerate, so choose what's best for you.

Kefir	Miso
Kimchi	Sauerkraut
Koji	Tempeh
Kombucha	Yoghurt

Kombucha shopping 101

We're seeing kombucha everywhere and it has made its way into the hippest and trendiest bars. However, not all kombuchas are created equal. Here is what you'll need to look out for:

- **Alcohol content:** Some kombucha brands contain more alcohol than is acceptable. Be safe and only consume kombucha where the brand states the alcohol content.
- **Sugar:** A small amount of sugar is absolutely fine, but beware since some commercial kombucha drinks contain so much sugar that you might as well be drinking a fizzy soda.
- **Artificial sweeteners:** We still have inconclusive data about what alternative sweeteners do to the gut, but I assume you're drinking kombucha as part of your gut health routine so be mindful of this.
- **SCOBY (symbiotic culture of bacteria and yeast):** Some mass-produced brands might be compromising the integrity of kombucha and its bacterial content by producing it from an extract, not from a SCOBY. Be on the lookout for strands floating in it – that's how you know it's the real deal.
- **If it's not refrigerated, don't buy it:** A low temperature is what keeps the yeast and bacteria alive.
- **Fake promises:** If the packaging makes a crazy claim (e.g. detoxifies, lowers blood pressure, anti-ageing) then STEER CLEAR, because none of that has been proven.

Onto more ingredients that nourish our gut microbiome:

The vivacious polyphenols

Think of polyphenols as the LGBT community of food. They are vibrant, powerful and simply fabulous. In simple terms, they are nutrients that are found in many plant-based foods and drinks. They're known for their antioxidant properties and, if you're new to the term 'antioxidant', an antioxidant is a vitamin, mineral or phytochemical (a naturally occurring plant substance) that protects our body's cells from damage caused by harmful molecules called free radicals. Some free radicals are made during normal body processes like breathing, exercising and digesting food. Others come from cigarette smoke and pollutants in air, water and food.

Polyphenols' work doesn't stop there. Around 5-10% of polyphenols are absorbed in the small intestine, while the rest travel down to the colon undigested to be broken down by our gut microbes. Polyphenols are antioxidants AND have prebiotic properties too. They've been shown to promote the growth of probiotic bacteria like Bifidobacterium and Lactobacillus and also to inhibit the growth of potentially pathogenic bacteria. These are great sources, listed in order of their polyphenol content:

Blackberry	Raspberry	Prune	Pear
Black chokeberry	Dark chocolate	Red currant	Yellow onion
Black currant	Chestnut	Peach	Apricot
Black elderberry	Black tea	Green olive	Asparagus
Blueberry	Green tea	Red onion	Almond
Globe artichoke heads	Apple	Green grape	White wine
Strawberry	Hazelnut	Potato	Turmeric
Plum	Red wine	Shallot	Cloves
	Black grape	Red chicory	Cinnamon
	Black olive	Broccoli	
	Spinach	Nectarine	

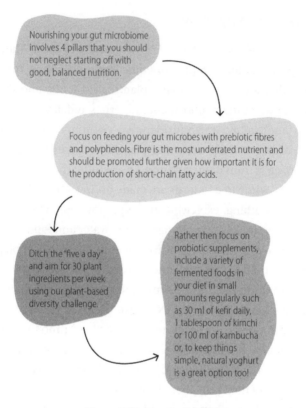

Nourishing your gut microbiome involves 4 pillars that you should not neglect starting off with good, balanced nutrition.

Focus on feeding your gut microbes with prebiotic fibres and polyphenols. Fibre is the most underrated nutrient and should be promoted further given how important it is for the production of short-chain fatty acids.

Ditch the "five a day" and aim for 30 plant ingredients per week using our plant-based diversity challenge.

Rather then focus on probiotic supplements, include a variety of fermented foods in your diet in small amounts regularly such as 30 ml of kefir daily, 1 tablespoon of kimchi or 100 ml of kambucha or, to keep things simple, natural yoghurt is a great option too!

Figure 5: Putting it all together

Where does '30' come from?

Thanks to findings by the American Gut Project, eating for a good gut has shifted from 'five-a-day' thinking to 30 plants per week. Researchers discovered that people who consumed 30 or more plant foods per week had a more diverse microbiome compared to those who ate 10 or fewer. They also found that those who ate 30+ plant foods had more microbes that produced SCFAs and you now know that's exactly what we want.

Including 30 plants a week in your diet may seem overwhelming, but here are some suggestions to get you started:

- Aim to work your way towards the following target from each group:
  - 13-15 different vegetables
  - four to five different varieties of fruit
  - three different pulses (beans, lentils, chickpeas, edamame)
  - four different grains and cereals (e.g. barley, whole-wheat couscous, brown rice)
  - five different nuts (three) and seeds (two)
  - a variety of herbs and spices (a minimum of six).
- Choose a three-bean mix that you can add to soups or salads.
- Go for a nut and seed mix that you can add to your cereals or salads.
- Try a new fruit and vegetable each week.
- Make friends with frozen food and choose a mix of frozen fruit and/or frozen vegetables.
- Think of the number 'three' – Add three different vegetables at lunch, dinner or both.
- Swap a meat dish with tofu, tempeh, beans or lentils.
- Add a new herb or spice every other day.

Plant-based eating FAQs

You've probably gathered that there's a lot of 'plant' talk when it comes to nourishing your inner ecosystem and I'm sure you've been bombarded by all things #plantbased or even #vegan. To clarify any confusion you may have about plant-based or plant-centred eating, here are the most common FAQs:

What exactly is a plant-based diet?

The name really says it all. A plant-based diet is one that is derived from plants and includes a well-planned variety of foods such as fruit, vegetables, nuts, seeds, legumes and wholegrain unprocessed products. It is also a diet that significantly reduces or eliminates the consumption of animal-based foods. Two keywords to keep in mind here are **unprocessed** and **plants**. Benefits of transitioning to a more-plant-based diet are endless – whether it's improving health or the environment, going plant-based will evidently make you feel good from the inside out if you've planned it well.

How does 'plant-based' differ from veganism?

Veganism involves strictly avoiding any animal-based products, including honey, as well as embracing an all-round lifestyle that considers animal welfare, avoiding leather and any other material made from animals. On the other hand, a plant-based diet can be seen as a big cloud that includes the different forms of vegetarianism with differing levels of eliminating animal foods.

Can I still call myself plant-based if I consume animal-based foods?

Frankly, I am no fan of placing labels on the way we eat as I am a firm believer that everyone can create their own template of eating that works best for them. Taking the best of what all the different diets have to offer and creating your own is one way to go. The term 'plant-based' actually doesn't confine you to one way of eating, which is what I like. In other words, you can still consider your diet plant-based, even if you do consume some animal-based foods, as long as the majority of your nutrition is derived from plants.

What are the best sources of plant-based protein?

Protein is generally one of the first nutrients people get concerned about when moving away from meat eating. In fact, we almost all exceed our daily protein requirements. Protein requirements range from 0.8 to 1.2 grams per kilogram body weight per day, and can even go higher in certain people such as athletes or those recovering from surgery. To give you a basic estimate of your individual requirement in grams (g), multiply your body weight (in kilograms – kg) by 0.8. For example, a person weighing 60 kg will require 48 g of protein per day as a minimum.

Plant-based sources of protein are abundant and include, but are not limited to:

- Legumes – beans, lentils, chickpeas: These are sources of both healthy carbohydrates and protein. They are also a great source of iron and fibre.
- Grains: quinoa, amaranth, buckwheat, wholegrain products.
- Nuts and seeds: chia, sunflower, pumpkin, flaxseed, linseed, walnuts, almonds, pistachios and organic nut butters, sesame, tahini.
- Tempeh, tofu and seitan. (Also known as 'wheat meat', seitan is based on gluten, the protein found in wheat.)
- Spirulina.
- Extras: plant-based protein powders made from lupin, peas, brown rice, hemp.

What would be the nutrients of concern if I were to reduce my consumption of animal-based foods?

As you make the switch to a more plant-centred diet, you will need to ensure that you are getting enough of the following nutrients, which may require you to do some additional planning depending on the level of restriction:

calcium	selenium
iodine	vitamin B12
iron	vitamin D.
omega-3 fatty acids	

Supplementation may be required if your blood levels of these nutrients are suboptimal.

How could I start making the transition to a more plant-based diet?

One of the easiest ways to start is to set one day a week to go meat-free or animal-free. For example, I encourage my clients to follow the #meatlessmondays movement as a start. It may sound easy at first but you will need to plan and focus on finding the right substitutes for a well-balanced meal. Another option would be just to cut out dairy for a week and use alternatives based on soya, almonds or oats that are fortified with calcium and vitamin B12.

What damages our gut microbiome?

We've worked our way through nourishing our gut microbiome but did you know that there are lifestyle factors that can be damaging to our inner ecosystem? These include smoking, alcohol, sugars and saturated fats.

Smoking

Firstly, to get smoking out of the way, just stop. Smoking is no longer chic and I simply want you to quit. It is one of the most damaging habits to our overall wellbeing and, in relation to gut health, it is a major risk factor for developing conditions such as inflammatory bowel disease (IBD) and cancers of the digestive tract. Research has shown that smoking can negatively impact the diversity of our gut microbiome and promote dysbiosis, where levels of pathogenic microbial species increase and beneficial microbes decrease. If you're a smoker and are reading this book, here's what I would recommend:

- Don't overthink this. Quit. Cold-turkey! Throw out your final batch to pave your way towards gut wellness.

- Consider nicotine replacement therapy as a means of managing any withdrawal symptoms and discuss your options with your doctor.

- Remember that cravings will pass and are only short-lived, so take a few deep breaths, grab a drink of water or make yourself a calming tea such as chamomile or lemon balm, and occupy your mind with something else until the craving passes.

- Understand your triggers and have a backup a plan.

- Make sure you have the right support around you and communicate your needs.

- Reward yourself with the money saved!

Alcohol

Alcohol is next up and let's just say this: it's a 'double-edged' sword. There's absolutely nothing wrong in enjoying a glass or two of good, polyphenol-rich vino but, in excess, alcohol can be detrimental to our gut microbiome and our overall gut health, including our liver and nervous system.

When we consume alcohol, it is absorbed in different parts of our digestive tract (our mouth and small intestine) and, once it diffuses into our bloodstream, is metabolised in our liver, meaning our liver functions as a detoxification system to get rid of it. Alcohol is broken down into a by-product called acetaldehyde, which is a known 'cell- poison' or carcinogen that causes a state of inflammation. This is why chronic alcohol abuse is associated with cancers of the mouth, oesophagus, stomach and large bowel, as well as liver disease and IBD.

Research investigating alcohol abuse and its impact on our gut microbiome has uncovered a few potential mechanisms by which alcohol consumption causes havoc within our digestive system. Dysbiosis and bacterial overgrowth are now recognised consequences of chronic alcohol consumption, in turn causing an increase in the production of toxins that cause inflammation. Alcohol can also weaken our gut lining, making it more permeable (i.e. 'leaky'). This can cause food particles to cross the gut lining, entering our bloodstream and triggering a cascade of immune responses, which is simply not a good thing. Is alcohol an issue you struggle with? Then here are a few pointers:

- The saying 'drink responsibly' goes a long way and my top tips would be to:
  o avoid quenching your thirst with alcohol
  o choose water instead first.

- When arriving at an event or a night out, get into the habit of choosing something non-alcoholic first then have your first drink after.

- Alternate between an alcoholic drink and a glass of water.

- Always have a balanced meal beforehand and avoid drinking on an empty stomach.

- Some of my favourite drinks to choose from include cucumber- or lavender-flavoured tonic water, ginger-based drinks and kombucha (see page 48).

- If you have a strong and uncontrollable desire to drink, then please do seek professional support, which is out there.

Refined sugars

Firstly, not all sugars are created equal! When I refer to the damaging effects of excess sugar, I am specifically talking about refined (added) sugars. Natural sugars that are found in fruit and dairy products (yoghurt and milk), and those that are broken down after consuming complex carbohydrates such as wholegrains and cereals, are not the ones we should be eliminating. It is a known fact that added sugars have been a health menace for many years and, thanks to the over-consumption and promotion of processed foods, like packaged baked goods, cereals and foods based on white flour, our health has taken a continuous dive into an inflammatory state. In relation to gut health, excessive consumption of added sugars has been seen to promote the growth of inflammatory microbes, such as *Akkermansia muciniphila* and *Bacteroides fragilis*, which may have a role in the development of conditions such as IBD.

We also see that excessive sugar consumption disrupts our inner ecosystem's harmony, causing dysbiosis, and while we still have a lot to uncover in this area, perhaps one thing that all gut health experts can agree on is that added sugar in excess damages our gut microbiome and overall gut function.

Saturated fat

Saturated fat is a nutrient that has been in the limelight for many years and the extent to which it is damaging remains controversial. We know that a high intake increases our risk of heart disease, mainly due to the impact it has on cholesterol levels. Research has also established that consuming too much of it plays a role in inflammation. Nowadays, the bad press continues when it comes to gut health; how so?

The three main consequences of consuming too much saturated fat are said to involve:

- the negative impact on microbial diversity
- increased intestinal permeability
- low-grade inflammation, ultimately affecting how our immune system functions.

Artificial sweeteners

Another controversial area of research lies in the world of artificial sweeteners. As I sit here writing this section, I am casually seated in the corner of a café, typing away. Close by my eye catches a couple popping sweeteners into their lattes. How appropriately timed, if I may say so, and I am not making this up...

The dieting industry and low-carbohydrate craze have given way to the world of 'sugar-free' foods with a heavy reliance on sweeteners. The couple sitting across from me are the perfect example of so-called health-conscious, possibly low-carb followers. The 'fight against obesity' and diabetes has led to dependence on sweeteners with the aim of allowing indulgence while avoiding waist expansion and blood sugar spikes.

To get some definitions out of the way, here is a quick look at the different types of sweetener currently used:

Artificial sweeteners are sugar-substitutes that are created in the lab and mimic the taste of sugar without the calories. Examples of these are acesulfame K, aspartame, cyclamate, saccharin and sucralose. These sweeteners are typically found in processed foods such as baked goods, soft drinks, canned foods, sweets, jams, dairy products and basically most products labelled as 'high-protein, low-carb'.

Natural no-sugar sweeteners include the popular stevia. Funnily enough, the commercial stevia that you end up buying off the shelves barely contains any whole-leaf stevia at all, putting the term 'natural' up for question. Such products are made from a highly processed stevia leaf extract called Reb-A so unless you're growing stevia at home, there's not much natural left. Stevia is popular amongst a number of hipster-looking brands that sell iced teas, protein bars and shakes, and bakeries claiming to look after your health.

Sugar alcohols are carbohydrates that are naturally found in some fruit and vegetables but can also be manufactured in a lab. They do contain calories but still in lower quantities than table sugar, hence their being another popular alternative. Sugar alcohols are found in chocolate, lollies, chewing gum and even toothpaste. No, they do not contain any ethanol despite the 'alcohol' part. Sugar alcohols are easy to identify as they end in '-ol' and include sorbitol, mannitol, maltitol, lactitol, erythritol and xylitol. You'll notice that foods containing sugar alcohols come with a warning that 'excess consumption may have a laxative effect'. Given that they travel all the way to our large bowel and are poorly absorbed there, excess consumption can cause bloating, stomach pain, gas and diarrhoea.

Researchers have started to unravel the long-term consequences of using sweeteners and controversy continues about the actual effects of their use on human health. Despite their apparent safety and extensive use, we have started to question whether long-term use of sweeteners truly offers any benefit in helping millions manage their weight, fight diabetes and protect their heart health.

When it comes to gut health, we're not fully there yet since a large chunk of findings comes from research on animals and we lack good quality studies in humans. Here's what we do know:

- Sweeteners may promote the growth of microbes that increase the bioavailability of calories.

- Sweeteners can influence the pathogenicity of gut microbes.

- Artificial sweeteners can change the composition and abundance of several gut microbial residents, potentially leading to dysbiosis.

Before ending this chapter, I want to remind you all that, despite highlighting individual nutrients that either nourish or damage our inner ecosystem, you need to look at your diet as a whole. As dull as the word 'balance' may seem, it truly does come down to balance and diversity, reminding us that nutrients work in synergy. Before eliminating foods, think about what you can include in your diet first that can give it that missing 'oomph'.

The good gut shopping list: A basic start to a gut-friendly kitchen

Pantry essentials:

- 70-80% dark chocolate
- Bag of mixed seeds
- Barley
- Brown lentils
- Brown rice
- Canned 2-3 bean mix
- Canned borlotti beans
- Canned chickpeas
- Canned corn
- Cashews
- Dates
- Dried figs
- Ground flaxseed
- Legume-based pasta
- Oats
- Sunflower seeds
- Walnuts
- Whole-wheat couscous

Fresh vegetables:

- Asparagus
- Aubergine/eggplant
- Beetroot
- Broccoli
- Carrots
- Cucumber
- Green beans
- Mushrooms
- Onions
- Spinach
- Sweet potatoes
- Swiss chard
- Tomatoes

Fresh fruit:

- Apples
- Berries in all their forms
- Citrus fruit
- Kiwi fruit
- Plums

Freezer essentials:

- 3-veg mix
- Frozen berries
- Frozen de-shelled edamame
- Frozen spinach

Herbs and spices (fresh or dried):

- Basil
- Cinnamon
- Ginger
- Nutmeg
- Oregano
- Rosemary
- Thyme
- Turmeric

Fridge essentials:

- Dairy alternatives such as soya or oat-based milks and yoghurts
- Green olives
- Natural Greek yoghurt
- Parmesan
- Tofu

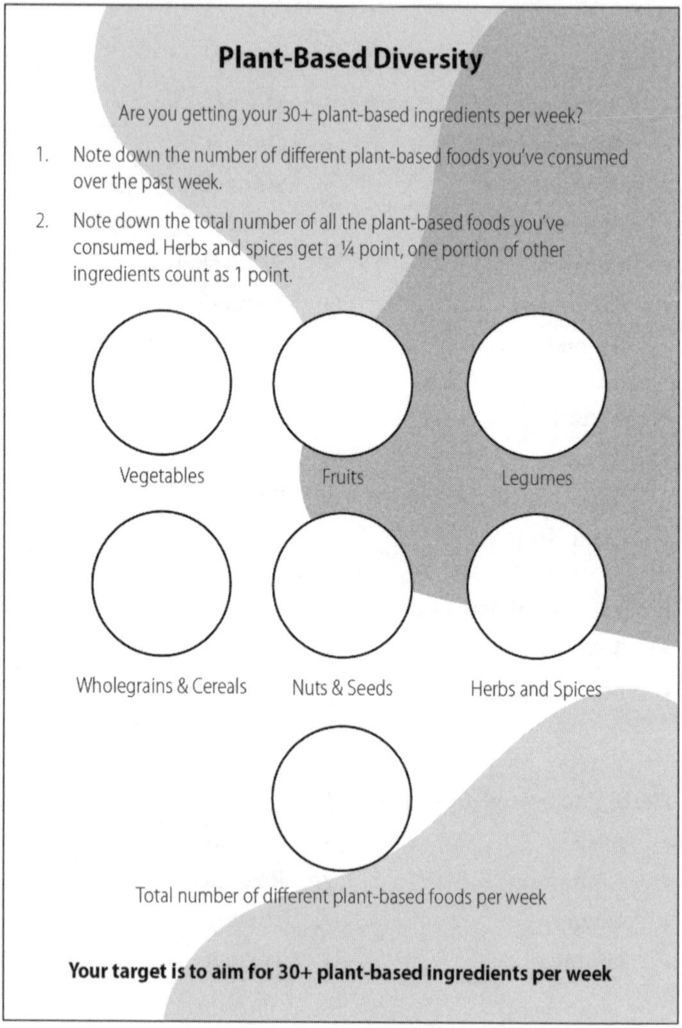

Plant-Based Diversity

Are you getting your 30+ plant-based ingredients per week?

1. Note down the number of different plant-based foods you've consumed over the past week.

2. Note down the total number of all the plant-based foods you've consumed. Herbs and spices get a ¼ point, one portion of other ingredients count as 1 point.

Vegetables

Fruits

Legumes

Wholegrains & Cereals

Nuts & Seeds

Herbs and Spices

Total number of different plant-based foods per week

Your target is to aim for 30+ plant-based ingredients per week

Figure 6: Plant-based diversity assessment

5

The one on GORD: the volcanic throat

Lara landed her dream job at a digital agency, leading a team to handle one of the company's most demanding accounts. This was no typical nine-to-five position; she had to manage the ins and outs of online marketing for one of her biggest clients to date. Work boundaries no longer existed thanks to her catering to different time zones, trying to keep her team above water to meet their client's deadlines and simply catching up on life. A year and a half into this position, Lara began to pay the psychological price for the chronically high stress levels she'd been living with and which had impacted multiple aspects of her life. Eating out for most meals was a norm, caffeine was neither friend nor foe but a necessity for a functioning mind, and broken sleep had become a common nightly occurrence. She started noticing a persistent cough that worsened around bedtime and when she woke up in the morning. On her bad days, the cough was accompanied by nausea, some chest pain and even a loss of appetite, which was her red flag that something was completely off. A few healthcare visits later, she was diagnosed with GORD and approaching burnout...

Earlier, when we spoke about the digestive tract, we talked through how food travels down the oesophagus. Now my friends, this chapter is all about what happens when food travels back up the oesophagus, and it is not pleasant. Firstly, let's talk about the terms 'heartburn' and 'acid reflux', which are used interchangeably but are actually very different things.

What is heartburn?

The name 'heartburn' is a little misleading, because the heart has nothing to do with it. Heartburn actually happens in your digestive system, specifically in your oesophagus. The lining of your oesophagus is more sensitive than your stomach lining, so any acid that enters it from the stomach can cause a severe burning sensation in your chest. This symptom usually occurs after eating and is quite common.

What is acid reflux?

The lower oesophageal sphincter or LES (a circular muscle that separates your oesophagus and stomach – see Figure 2) is in charge of closing off your oesophagus after food has passed to your stomach. If the muscle is weak or doesn't tighten properly, the acid from your stomach can move backward into your oesophagus. This is known as acid reflux. Acid reflux can cause heartburn, a cough, a sore throat and a sour taste in the mouth.

What if this keeps occurring? Then we're looking at gastro-oesophageal reflux disease (GORD). As defined by the Gastroenterological Society of Australia (GESA), for example, GORD is a condition where the ongoing reflux places the patient at risk of complications or symptoms that can impact their quality of life. It is also more common than we think.

GORD is seen as a collection of symptoms due to the reflux of gastric contents into the oesophagus, larynx, mouth and/or lungs. So GESA tells us that GORD (or GERD depending on where you come from) is common and can affect our quality of life. So, what triggers it? The answer is not that straightforward as the condition is multifactorial and complex, but it involves factors that affect the functioning of your LES causing it to become lax. These causes include:

- Increased pressure on the stomach: For example, during pregnancy; after eating large amounts of food and fluid at the same time; wearing tight-fitting clothes.

- Increased acid secretion in the stomach and a slower rate of stomach emptying: This can be due to eating meals that are high in fat and protein.

- Consuming fat, alcohol and/or coffee, as well as smoking, which may all lead to the relaxation of the valve at the opening of the stomach. They also cause increased acid secretion in the stomach, with reduced LOS pressure as the common factor.

- Medications: Some medication may be responsible for GORD, such as anti-inflammatory drugs (e.g. ibuprofen), antibiotics and some antidepressants. When in doubt, always consult your doctor if you feel like your medication may be the culprit.

What about stress?

We are currently living in a society where anxiety, burnout and being overworked have become a widespread struggle-combo and a social-norm. Add to that the effects of the pandemic, causing a tsunami of mental health catastrophes. Working from home has caused 'life and work' to merge and boundaries to become non-existent; burnout has become the new pandemic. This observation has no conclusive, scientific backing, but seeing a huge influx of clients into my clinic with reflux over the past couple of years has led me to look into stress and stomach acid levels. The research remains inconclusive as to whether stress causes an increase in stomach acidity, but scientists believe that we become more sensitive to acidity in the oesophagus when we're stressed.

When we're stressed, EVERYTHING is amplified. Researchers hypothesise that changes in our brain chemistry when we're stressed are responsible for making us more sensitive to minor increases in acidity. The big 'S' can also cause a reduction in compounds called prostaglandins that protect our stomach from the impact of acid, in turn, amplifying discomfort and overall symptoms.

Signs of GORD

This brings us to the symptoms of GORD. The two most obvious to watch out for are reflux and heartburn as discussed already, but there are a number of other red flags that may indicate a case of GORD:

- Chest pain, dysphagia (difficulty swallowing), belching
- Epigastric pain
- Nausea, and bloating
- Cough, voice hoarseness
- Throat pain or burning
- Sleep disturbances.

What should you do if you have GORD?

Treatment generally ranges from lifestyle measures to medical intervention and, in extreme cases, surgery.

Lifestyle changes remain the cornerstone of any therapeutic approach to managing GORD, yet we are still devoid of high-quality trials providing clear evidence for avoiding specific foods or changing certain habits. By saying that though, before jumping on any medical intervention, all researchers agree that it is worth giving lifestyle and diet a go first.

Temporarily avoid these

Let's look at foods to limit or temporarily avoid. The rationale for avoiding these dietary triggers is based on the theory that some foods relax the LES, allowing stomach contents to back up into the oesophagus. The list includes:

- Alcohol
- Chocolate
- Citrus fruit
- Cola, coke, coffee, cocoa, chocolate
- French fries, potato crisps
- Fried eggs, fried or battered meat, bacon, sausages, salami, pepperoni, hot dogs
- Full-fat milk, cream, yoghurts, chocolate milk, cheddar cheese and ice cream
- Gravy
- Pastries and doughnuts
- Peppermint
- Spices such as chilli, curry mixtures, garlic, black pepper.

While mentioning the foods listed above, I just want to note that it is unlikely that all these foods would trigger your symptoms.

If you want to identify whether a particular food is making your symptoms worse, keep a detailed food and symptom diary such as the one below. We will be using this diary with other conditions too to identify symptom triggers.

Date: **Daily Food Diary**

Symptoms/feelings i.e. cramps, bloated

Breakfast:

Lunch:

Dinner:

Snacks:

Movement:

Stool (circle me)

Water (tick me) ☐☐☐☐☐☐☐☐☐☐   Or litres per day: ☐

Stress (circle me)   1   2   3   4   5   6   7   8   9   10

Figure 7: Your food diary

Eating behaviours to consider

Alongside nutritional changes, you can't neglect behaviours and habits so try these too:

- Sit up straight while eating.

- Posture training seems to have helped some of my clients so there's no harm in working on your slouch.

- Eat slowly and chew foods properly. If you are regurgitating large chunks of food, that is your gut telling you to slow down and chew.

- Avoid drinking fluids with meals and instead try drinking 30 minutes before or after eating.

- Do not go straight to bed after a meal. Have your meal two to three hours before bedtime.

- Stop smoking.

What about sleep?

Medical guidelines are finally highlighting the importance of other lifestyle pillars when it comes to managing gut conditions and, in a recent review, patients with GORD were advised to improve their sleep hygiene. Why? According to researchers, sleep reduces GORD by suppressing a mechanism called transient lower oesophageal sphincter relaxations (TLESRs). These relaxations are the main mechanism of GORD (but not the cause) and happen spontaneously, independently of swallowing.

In Chapter 12, we will talk about sleep in detail and I highly recommend applying the 'Sleep Hygiene' guidelines on page 218 whether you suffer from GORD or not. Once again, this is further confirmation that managing gut conditions will require you to address multiple lifestyle pillars that work in synergy.

Let's briefly address medication...

If you've been diagnosed with GORD, there's a high chance that you've been prescribed medication to manage it, if lifestyle measures have failed. I am fully aware that many doctors are quick to jump on their prescription notepads, but a few lines down, you will realise that it is important to create dialogue with your healthcare professional on what the best and safest approach would be should medical therapy be considered.

The types of drug often given for reflux include the following: antacids, histamine 2 receptor antagonists (H2RA) and/or proton-pump inhibitors (PPIs).

Antacids: Over-the-counter-medication that 'neutralises' stomach acid in turn, providing relief from indigestion and heartburn. You do not need a prescription and they come in either liquid or chewable tablet form. Popular brands include Gaviscon, Pepto Bismol, Tums and Alka-Seltzer.

H2RA: These meds are also called 'H2 blockers'. They were the first drugs used to effectively treat stomach ulcers and GORD. As the name suggests, they block stomach acid production by competing with histamine to attach to their respective docking stations – that is, H2 receptors. Histamine stimulates the cells in your gut lining to produce stomach acid aka hydrochloric acid. By H2RA blocking histamine, the cells in the lining of your stomach will receive less stimulation to produce acid, in turn, offering relief.

PPIs: Proton pump inhibitors have been considered the most effective therapeutic approach to manage GORD and are widely used due to their consistency in suppressing acid production. Medical therapy using these drugs involves taking them for a course of four to eight weeks with an adjustment of dosage if needed. However, there is a downside... They were considered safe for many years until the last decade unravelled a flood of studies highlighting multiple side effects of long-term treatment. These now recognised side effects are:

- Nutrient deficiencies such as magnesium and vitamin B12 due to the decreased acid secretion, which impacts their absorption
- *C. difficile*-associated diarrhoea
- Osteoporosis and high risk of fractures
- Negative effects on gut microbiota contributing to dysbiosis.

For this reason, it is evident that PPIs should be recommended only when strictly necessary. The medical community (well, the one that I am a part of) is in agreement that lifestyle measures are foundational to managing GORD and that those requiring medication should receive the lowest dose to control their symptoms. This brings us to the question of whether alternative therapies can help.

Can alternative therapies help?

With all the negative press around long-term PPI use, many have started to resort to alternative therapies and the most popular ones currently trending include slippery elm, curcumin, melatonin and aloe vera juice. The main mechanisms reported for most alternative supplements include offering protection against inflammation, inhibiting acid secretion or rebuilding the lining of the oesophagus. Unfortunately, the science is lacking or very weak when it comes to supporting their use (test tube and animal studies are not strong enough for us to safely recommend them). However, if there's one alternative approach that has been extensively researched in GORD management, it's quince syrup.

Quince

I first came across quince syrup when my first-born was struggling with horrific reflux. Given that I am connected with some excellent, integrative gastroenterologists, I naturally reached out to them for advice and the recurring answer was quince syrup. Quince is a fruit commonly grown in south-eastern Europe and Latin America and resembles a cross between a lumpy apple and a pear. Quince syrup is believed to help manage symptoms of GORD just as well as medication but mainly in children and pregnant women. The exact mechanism remains unclear as we are unsure whether the positive outcomes have to do with compounds in quince impacting muscular contractions, reducing acid secretion or alleviating the discomfort associated with reflux. Nevertheless, we did see a difference and it has become an essential addition to our pantry. In terms of the dosage, it varies depending on age, but for older children and adults, 10-15 ml once a day for six weeks could be a reasonable place to start.

Remember though that, just because something is natural, it doesn't mean it's always safe. You also always need to keep in mind that natural compounds in food can interact with medication so, for example, quince syrup may interfere with the absorption of certain drugs, rendering them less efficient. Before jumping on the alternative therapy bandwagon, please consult your healthcare specialist and create dialogue around why you'd like to give alternative therapies a go and whether it would be a safe option for you.

6

The one that's all about the bloat

It's Thursday afternoon, post lunch, and staff at G&J are rushing into the most unpleasantly set up conference room. Since this book was being written in the midst of a pandemic, let's assume they're all being responsible and have masks on. Jay, a senior executive, rushes in with a pile of reports and Leanne, his assistant, is not far behind. Laurence, a new intern, is eagerly waiting to listen in on her first meeting about the company's budget and financial forecast.

Three very different humans with one very common problem. As Jay sets up his presentation to kick off their meeting, he is struggling to ignore the fact that he's about to 'let one rip' and is seriously regretting the bowl of Brussels sprouts and cabbage salad he so excitedly thought was a healthy addition to his lunch. Leanne is standing in a corner, too concerned that colleagues may think she's well into an unannounced pregnancy. And Laurence, poor Laurence has broken into a sweat and feels like a storm is brewing in her gut, and she too could 'let a mighty one explode.'

Bloating is such a normal physiological response to digestion yet can be so disruptive, painful and, put simply, a nuisance. It is one of the most common digestive complaints in people of all ages. Many express feeling 'pregnant', 'being too gassy', 'feeling heavy' and 'uncomfortable in the stomach' with or without a visible increase in stomach girth. The funny thing is, despite being one of the most frequent and bothersome complaints, we still don't fully understand the exact cause of bloating since it's not that straight forward. Wondering why? Behold (Figure 8) the possible reasons for bloating and that's excluding pathological causes.

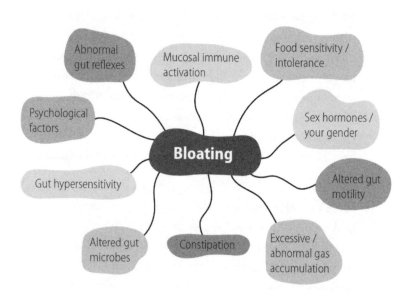

Figure 8: Possible reasons for bloat

I've dedicated specific chapters in this book to a few individual causes, such as constipation, food intolerance and irritable bowel syndrome (IBS), but this chapter serves as a starting point. So, going back to definitions, bloating is a sensation, abdominal distension is a sign. Bloating is a subjective sensation of feeling 'gassy', 'having trapped gas', 'pressure' without visible distension, while abdominal distension refers to a visible increase in abdominal girth.

If we look at organic – that is, pathological – causes of chronic bloating, the list is quite extensive and will require your doctor to rule out a number including:

- SIBO – small intestinal bacterial overgrowth
- Lactose, fructose, and other carbohydrate intolerances
- Coeliac disease
- Pancreatic insufficiency
- Gastric outlet obstruction
- Gastroparesis
- Ascites (fluid retention in the abdominal cavity)
- Gastrointestinal or gynaecologic malignancy
- Hypothyroidism
- Small intestine diverticulosis
- Chronic intestinal pseudo-obstruction.

Before any of these big words causes alarm, your doctor will run the necessary tests that make the most sense in line with your medical and symptom history, so I truly doubt that a case of the bloat is an intestinal obstruction or malignancy. The clinical relevance of bloating as a symptom is extremely variable.

Firstly, we bloat. It's a fact of life and we're not meant to be walking around with wash-board abs all the time.

What is a 'normal bloat'?

One way I help my clients distinguish a 'normal' bloat from a worrisome one is by using the following guide:

- Using a scale from 0 to 10, where 0 = no discomfort and 10 = excruciatingly uncomfortable, rate your level of discomfort.

- A score of 0 to 5 tends to reflect normal processes of digestion and bloating as a result of either eating too fast, too much or just a lot of fermentable food and fibre.

- A score of 6 to 10 is worth looking into and asking for help from a dietitian. You should certainly seek help if you rate your discomfort consistently over 5. I would suggest consulting a specialist if the discomfort is also accompanied by stomach pain that is not relieved when you 'do the deed' or toot the gas away, as well as noticing any changes in your bowel movement.

How we bloat

The mechanics of bloating are not that straight forward, but simply speaking, the increased pressure in our intestines is a result of the volume of food and fluid consumed, or of gas produced by our gut microbes when we've overfed them with fibre and fermentable carbohydrates. The increase of volume in our gut stretches our intestine, leaving us with the sensation of 'bloating'. Some people are more sensitive than others when it comes to this intestinal stretch, such as those with living with IBS. This is called visceral hypersensitivity. Furthermore, the bloat can be overly exaggerated in those living with functional gut disorders where the mechanism behind 'looking six months pregnant' involves the improper movement of the diaphragm and the belly causing the stomach to protrude. This is called an abnormal viscerosomatic reflex that is triggered due to stomach pain and discomfort. This reflex is an example of how our gut and brain communicate – the gut sends an SOS to the brain and in turn, the brain contracts the diaphragm and relaxes the muscles in the abdomen, causing the belly protrusion.

Let's talk about gas for a minute

Gas in our intestines ends up there from two main sources – the air we breathe and the fermentation of food in our intestines, i.e. our gut microbes breaking down and fermenting food in the large intestine. In healthy humans, 70% of the total gas present in the gut is located in the large intestine. About 20% of the gas produced there is expelled via the anus and the rest is eliminated via other routes – for example, being reabsorbed in the small intestine or by colonic bacteria.

Flatulence (also known as tooting, trumping or farting) is the formal name for releasing gas from the digestive system through the anus. Did you know that most people 'toot' about 10-25 times a day? While totally normal, gas can be embarrassing and can sometimes cause uncomfortable bloating, pain or discomfort. If you're passing wind more than 10 times a day, figure out what part of the process is bothersome. Is it the smell? Did you soil yourself? (Oh, that is actually a lot more common than you might think and is also known as 'wet-farts', my friends... .) Not being able to hold it in?

Toots are not meant to smell like poo-pourri and lillies, but if you notice a stench of rotten eggs and the odour of your flatus is bothersome, the culprit could very well be the food you're consuming. The distinct smell is a result of your gut microbes breaking down foods that contain sulphur. Foods containing sulphur include:

- Meat, chicken, eggs, protein supplements, cold cuts
- Vegetables such as broccoli, cabbage, kale, cauliflower, Brussels sprouts
- Allium vegetables such as onion, leeks, garlic and chives
- Beer and wine.

When you consume these foods, your gut microbes will break down the sulphur-containing compounds producing sulphur-containing gases such as hydrogen sulphide. One common complaint from clients who have been on very high protein diets or are heavy beer drinkers is the funky smelling flatus. My suggestion? Before, eliminating vegetable culprits, start with addressing the excess protein first and the alcohol.

Solutions for excess gas

If excess gas is a problem, take note of these pointers:

1. How to swallow less air:
   - Chew food well, with your mouth closed.
   - Do not use a straw to drink.
   - Avoid chewing gum.
   - Avoid carbonated drinks and sparkling water.
   - Try not to overeat and stop when you reach the sensation of 'I am satisfied'.
   - Drink lukewarm fluids instead of hot beverages.

2. Temporarily limit gas-causing foods. You will need to identify your individual triggers but the goal is not about total elimination; rather it is about strategically eating them in smaller amounts and not combining them together. The list includes:
   - Apples and apple juice
   - Artichokes
   - Asparagus
   - Beans
   - Bok choy
   - Broccoli

- Brussels sprouts
- Cabbage
- Cauliflower
- Lentils
- Lettuce
- Onions
- Peaches
- Pears and pear juice.

For example, if you want to consume legumes such as lentils, avoid combining them with onion and/or broccoli. A good start would be ½ cup of cooked lentils with spinach, carrots and courgettes (zucchini). Tip: If legumes are an issue, use canned legumes; drain and rinse them well first before eating.

Managing the bloat

When it comes to therapeutic approaches to managing bothersome bloating, knowing the cause is obviously key. However, whether you're dealing with an organic cause or a functional problem, your management plan may include one or more of the target areas shown in Figure 9.

Eating behaviours: where to start?

Before jumping on any elimination bandwagon, don't forget to start with tweaking certain behaviours:

- If you're a fast eater, slow down! The target to aim for is 15-20 minutes to finish a meal.
- Chew your food well.

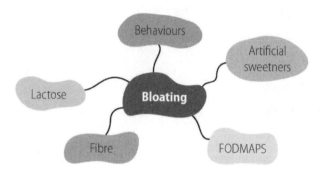

Figure 9: Possible casues of bloating

- Avoid consuming large meals and overeating. Have smaller, more frequent meals such as three mains in a day and one snack in between lunch and dinner.

- As I've mentioned above, carbonated drinks and chewing gum cause you to swallow more air, contributing to the bloat.

- Loosen up! Skinny jeans and tight yoga pants are the worst choice for bloating. Wearing tight pants constricts your stomach due to the extra pressure, making it more difficult for food and gas to move along. Also, as your tummy relaxes and contracts, tight clothes can make this pressure more noticeable and definitely unpleasant. So, feeling bloated often? Skip the skinnies and opt for more comfortable choices.

Fibre: Too much, too little, just right?

We now know that fibre is a crucial ingredient for good gut health but it is definitely a known trigger for bloating. Many of my clients who have made the switch to a more plant-centred diet have complained of excessive bloating. In this case, bloating is a normal physiological response to fibre being fermented in their large intestine. The mistake though, is increasing fibre too quickly; that is a perfect brew for a stomach catastrophe. To put things into perspective, a regular Western diet provides us with an average of 15-20 grams of fibre per day. Going from 15 to 30-40 grams overnight (which is what a plant-based diet generally provides) will cause excessive gas, stomach pain, bloating and, in some cases, diarrhoea or even constipation. Should that put you off including more plants? Absolutely not! The trick?

- Gradually increase your fibre intake over four weeks.
- Make sure you drink enough water to prevent constipation, which can ultimately make the bloating worse.
- Spread your fibre intake throughout the day.
- For some, cooked vegetables in the evening may be tolerated better than raw, salad vegetables.

Sandra's four-week plan for increasing fibre

Week 1
- Add ¼ cup of canned chickpeas to some mains three days a week
- Include 1 tbsp of seeds to your breakfast or snack
- Make sure you consume 1.5-2 litres of water daily.

Week 2
- Maintain changes in week 1
- Ensure you include three different types of vegetable at lunch
- Add ¼ cup of nuts to your snack daily
- Increase your water intake to 2 litres daily.

Week 3

- Maintain changes made in weeks 1 and 2
- Add three different vegetables at dinner
- Try four different types of fruit this week e.g. kiwi fruit, strawberries, frozen berries and bananas
- Increase your water intake to 2.5 litres.

Week 4

- Maintain changes to date
- Challenge yourself to include 30 different plants this week using the diversity guide in Chapter 4
- Maintain fluid intake between 2 and 2.5 litres.

The first line of elimination: Artificial sweeteners

We spoke about artificial sweeteners in Chapter 4. The ones that you can eliminate to reduce bloating, if they are a common in your diet, are the sugar alcohols. These include sorbitol, mannitol, maltitol, lactitol, erythritol and xylitol. As I mentioned previously, sugar alcohols travel all the way to your large bowel and are poorly absorbed there; consequently, if consumed in large amounts they can cause bloating, stomach pain, gas and diarrhoea.

Food sources of sugar alcohols include the following and I would temporarily avoid them:

- Low-carb, high-protein dairy products, such as high-protein milks, puddings, yoghurts and desserts.
- Protein shakes, mainly the ones that are flavoured.

- Sugar-free gums and lollies.
- 'Diet', sugar-free foods.

Could it be lactose?

Another very common cause of bloating is lactose intolerance. Lactose is the sugar found in milk and dairy products. To break down lactose, we need an enzyme called lactase and with lactose intolerance, our bodies do not make enough lactase. Lactose then travels undigested to the large intestine where it is broken down by our gut microbes and the malabsorbed lactose causes symptoms such as bloating, abdominal pain, gas and diarrhoea within 30 to 120 minutes of ingestion.

There are different types of lactose intolerance and the two most common types are:

- Primary lactase deficiency: This is the most common type where we produce less lactase in adulthood as we become less reliant on milk and dairy products in comparison to childhood.
- Secondary lactase deficiency: This type may be temporary, due to a medical condition such as a gut infection, active coeliac disease or inflammatory bowel disease.

If you do suspect that lactose may be a culprit of your bloating, then a temporary elimination for two weeks would be advised, and if your symptoms improve, then lactose intolerance could be the cause of the uncomfortable bloat.

How to eliminate lactose

What should I eliminate for two weeks?

- Cow's milk
- Goat's milk

- Some types of cheese, such as cream cheese, mascarpone, ricotta, mozzarella
- Cream
- Some yoghurts
- Ice cream
- Foods containing the following in the ingredients' list:
  - Milk
  - Milk solids
  - Malted milk
  - Whey (whey protein concentrate has more lactose than whey protein isolate and whey protein hydrolysate)
  - Lactose
  - Curds
  - Cheese flavour
  - Cream
  - Yoghurt
  - Buttermilk
  - Sour cream
  - Non-fat milk solids
  - Non-fat milk powder.

The following ingredients do not contain any lactose:
- Lactalbumin
- Lactate
- Lactic acid
- Casein (which is the protein found in milk).

People with lactose intolerance are generally able to tolerate small amounts of lactose ranging between 12 and 15 grams at a time. This is equivalent to about 1 cup of milk. Most can tolerate up to 24 grams of lactose per day if it is spread out strategically throughout the day.

What can I eat on a lactose-free diet?

Luckily, we have become spoiled for choice as lactose-free products are readily available now, including lactose-free milk and yoghurt, but here's a list of low- and no-lactose foods:

- Hard cheese such as Parmesan, gouda and edam
- Kefir
- Yoghurt with live cultures
- Low-lactose cheese such as cottage and feta
- Plant-based milks, yoghurts and spreads such as soya, almond, coconut, oat, pea and rice, but make sure that these products are fortified with calcium (i.e. have calcium added).

You can also consider taking a lactase supplement, where the enzyme comes in supplement form as a pill or drops, which you can take right before consuming a meal with lactose in it. The dosage will depend on the brand and the estimated amount of lactose in your meal, so ask the pharmacist or your accredited nutritionist for guidance.

And finally, let's get acquainted with FODMAPs – but not too acquainted just yet as I'll be talking about the FODMAP process in greater detail when we get to Chapter 9 on irritable bowel syndrome (IBS).

If you're concerned about calcium, fear not!

Most dairy alternatives will contain added calcium so look out for calcium-fortified plant-based milks but also, don't forget about the following foods as they are a good source of calcium too:

- Vegetables such as collard greens, kale, broccoli, bok choy, spinach and turnip greens
- Almonds and tahini (sesame paste)
- Seeds such as sunflower and chia
- Canned salmon and sardines with bones
- Soya beans, tempeh and calcium-fortified soya products
- Navy beans.

A first glance at FODMAPs

FODMAPs is a term given to a group of poorly digested, rapidly fermentable carbohydrate food molecules. The term FODMAPs stands for fermentable oligo-, di- and mono-saccharides and polyols. Examples of these include fructose (e.g. sugar found in fruit); fructans found in wheat and numerous vegetables such as onion; lactose (found in milk); and xylitol (artificial sweetener). If consumed in large amounts, FODMAPs are believed to increase the volume of liquid and gas in the small and large intestine, resulting in abdominal pain, gas and bloating. Controlling the amount of FODMAPs in your diet is known to reduce symptoms.

The ability of our bodies to digest FODMAPs varies from person to person. People with specific gut disorders such as IBS are believed to experience more uncomfortable symptoms due to the gut being unusually sensitive.

Examples of high FODMAPs that would need to be temporarily restricted are listed in Table 2.

Table 2: Foods high in FODMAPs

Excess fructose	Fructans	Lactose	Galacto-oligosaccharides	Polyols
Apples	Artichokes	Custard	Beans (kidney beans, baked beans...)	Apples
High-fructose corn syrup	Asparagus	Dairy desserts	Chickpeas	Apricots
Honey	Beetroot	Ice cream	Lentils	Avocado
Mango	Garlic	Milk		Cherries
Pear	Leek	Milk powder		Lychees
Watermelon	Onion	Soft, unripened cheeses (e.g. ricotta, cream, cottage, mascarpone)		Mushrooms
	Radicchio			Nectarines
	Rye			Pears
	Wheat			Plums
				Sorbitol
				Xylitol

When clients embark on the popular low-FODMAP diet, examples of what they can consume include what is shown in Table 3.

Table 3: Foods low in FODMAPs

Vegetables	Fruit	Dairy products	Grains
Aubergine/ Eggplant	Bananas	Hard cheese	Gluten-free breads and cereals
Carrots	Blueberries	Lactose-free products	Polenta
Celery	Cantaloupe	Rice milk	Rice
Collard greens	Clementine		Quinoa
Cucumber	Grapes		Sourdough bread
Green beans	Kiwi fruit		
Pumpkin	Oranges		
Spinach (baby spinach)	Passion fruit		
Swiss chard	Pineapple		
	Strawberry		

It is important to note that eliminating FODMAPS is a three-phase process and the elimination phase is generally followed for two to six weeks to achieve symptom control. You will then enter the reintroduction phase. This phase will help you determine your threshold of tolerance and which FODMAPs you should continue to restrict. Phase 3 of the diet aims to liberalise your diet further while maintaining good symptom control.

I will be describing the FODMAP process in detail with a more comprehensive food guide when we talk about IBS (page 138) but I need you to know that the elimination phase comes with a ton of cons, which we will go through – compliance for one, but also depriving your inner-ecosystem of valuable fibres that help it thrive and the risk of developing disordered eating. Placing bloaters directly on a low-FODMAP diet has quickly become the default but we are now learning about the negative consequences of such actions and I urge any doctors reading this to work closely with an experienced dietitian to evaluate whether having a patient go through the FODMAP process is the ideal approach or whether less restrictive modifications can be made.

Probiotics and bloating

One of the common questions I get is whether probiotic supplements can be used to manage bloating. The simple answer is, not always! It may seem like an attractive option given that they have a role in modulating fermentation, inflammation, visceral sensitivity and pretty much everything that has to do with gut health, but unfortunately, it is way more complex than it appears at first glance.

Many studies looking at probiotic use to beat the bloat lack insight into the potential mechanism of how this is achieved. There are specific strains of probiotics that seem promising and they include:

- *Lactobacillus acidophilus* NCFM
- *Bifidobacterium lactis* Bi-07
- *Lactobacillus plantarum* Lp299v
- *Bifidobacterium infantis* 3562411

These strains seem to put the brakes on gas production, perhaps due to their restoring the balance of the gut microbiome to a healthier one and reducing gas-producing bacterial species and/or transforming how they function.

If you want to consider a course of probiotics to address your bloat, work with your health professional to ensure that you've chosen the right strain first for a period of four to 12 weeks.

What about activated charcoal for bloating?

Another popular remedy to beat the bloat that's made its rounds is activated charcoal. It is a type of charcoal that's been treated with heat to make it more porous. This form makes it more effective in trapping gas, subsequently alleviating the bloat. Activated charcoal has primarily been used for the emergency treatment of poisoning or drug overdose in a medical setting.

Activated charcoal has been sold in supplement form as well as powders and has been added to so-called juice cleanses but unfortunately research has been limited although promising. However, it should be taken only with the guidance of your specialist given the side effects that accompany its use. These include vomiting, constipation and black stool.

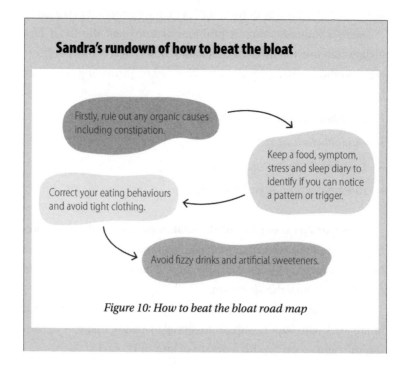

Sandra's rundown of how to beat the bloat

Firstly, rule out any organic causes including constipation.

Keep a food, symptom, stress and sleep diary to identify if you can notice a pattern or trigger.

Correct your eating behaviours and avoid tight clothing.

Avoid fizzy drinks and artificial sweeteners.

Figure 10: How to beat the bloat road map

1. Firstly, rule out any organic causes including constipation.

2. Keep a food, symptom, stress and sleep diary to see if you can notice a pattern or trigger.

3. Correct your eating behaviours and avoid tight clothing.

4. Avoid fizzy drinks and artificial sweeteners.

5. Limit some gas-causing foods temporarily, such as the ones mentioned in the section on gas:

- Apples and apple juice
- Artichokes
- Asparagus
- Beans
- Bok choy
- Broccoli
- Brussels sprouts
- Cabbage
- Cauliflower
- Lentils
- Lettuce
- Onions
- Peaches
- Pears and pear juice.

If the first few points do not offer any relief, you can consider a temporary elimination of lactose to see if that resolves your bloat.

If lactose is not the culprit, we can then look at a less restrictive version of the low-FODMAP diet for two weeks (see Chapter 9).

Consider probiotics only under the guidance of your dietitian.

7

The one for the faecally challenged

It's been three agonising days with her bowels showing no signs of movement or emptying. A heavily pregnant Gemma has been struggling with the worst constipation her bum has ever experienced during the final stretch of her pregnancy and it feels like she's completely lost faith in the power of prunes. Two floors down, Luke has been comfortably sitting on his loo for 20 minutes, flicking through this Sunday morning's paper. Deed done, he walks out still feeling like he's got more to empty and he can't help but feel he's had the most unsatisfying Sunday-morning poo. Across the street, Alissa is huddled on her couch in pain, bloated and confused. She has been frantically researching ways to relieve her constipation and goat-like stools and is weighing up the options of a DIY enema before her upcoming graduation.

Here's where the in-depth poo talk begins! These are all real-life tales of constipation struggles. Before we get into it, let's begin by breaking down everything you need to know about poo...

The Bristol 'poo' chart

If we're going to talk about poo you'll need to know about the Bristol Stool Chart if you haven't come across it already. This chart has been used to help people communicate their poo type. It classifies stool into seven different categories as per the image below:

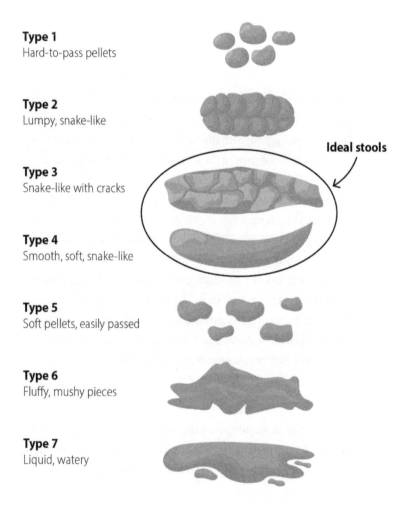

Type 1
Hard-to-pass pellets

Type 2
Lumpy, snake-like

Ideal stools

Type 3
Snake-like with cracks

Type 4
Smooth, soft, snake-like

Type 5
Soft pellets, easily passed

Type 6
Fluffy, mushy pieces

Type 7
Liquid, watery

Figure 11: The Bristol Stool (or 'poo') Chart

Types 1 and 2: These types indicate a tendency towards constipation, which we will be discussing in this chapter. These stools appear as separate, hard lumps or pebble-like (type 1) or snake-shaped but lumpy (type 2). These stools tend to be hard, dry and difficult to pass, since stool has been hanging out in your large intestine for a longer period of time. Clients who are type 1s and 2s tend to complain of incomplete evacuation, meaning that they are unable to pass any more stool despite feeling that their rectum is still full.

Types 3 and 4: When I talk about an ideal poo, I am referring to either a type 3 or type 4 and that's my mission when working with clients: getting them to rejoice when either is passed! A type 3 stool refers to a sausage like poo with cracks, whilst type 4, is a soft, smooth, snake-like poo. Both types are soft and easy to pass with little effort and, upon a good bum-wipe, you should see nothing or have minimal residue on the toilet paper.

Type 5: These are soft blobs with clear-cut edges that are easily passed. The consistency may lead many to think that they're experiencing diarrhoea but that's not the case.

Types 6 and 7: When we're experiencing either of these, we're talking about a case of loose poo that's not ideal. Type 6 is mushy poo and type 7 is watery with no solid pieces. Both types 6 and 7 create an urgency to go to the loo such as in the case of a stomach bug, food poisoning or a food sensitivity or allergy.

Let's start off with the first set of poo types that lean towards constipation.

What is a normal bowel movement?

You'd be surprised to hear that the answer is not that simple. We have said that the ideal would be a type 3 or 4 stool passed with ease. You will want to assess what your poo looks like (the structure, colour, etc.).

What is not normal?

- Pooing more than three times a day or less than three times a week warrants some investigation.

- When you have difficulties, such as straining and pain, it can be a potential sign of constipation.

- If you notice blood in your stool or your stool is really dark (tar-like), I recommend you contact your doctor immediately.

- Any change in your bowel movements (diarrhoea, or alternating between constipation and diarrhoea) accompanied with pain or bloating is also a sign that you might need to speak to a professional.

Defining constipation

Did you know that you can go to the loo daily and still be constipated? The definition of constipation has come a long way but it's also quite subjective. If we had to look at the formal description, it refers to having two or more of the following for more than 25% of defaecations:

- Straining
- Lumpy or hard stools
- Sensation of incomplete evacuation

- Sensation of anorectal obstruction or blockage
- Less than three bowel movements per week.

Also, many of you may not realise that there's more than one type of constipation.

Types of constipation

The first consideration is whether constipation is the primary problem or is secondary to another factor:

Primary constipation (PC), also known as functional constipation, is the most common form, where the constipation is the primary medical condition. Now, there are sub-types of PC:

- Normal transit constipation: The person thinks that they are constipated, but the transit and consistency of the stool are normal.
- Slow transit constipation: This means that the transit of the stool is slower, due to slower stimulation in the bowel movement, also referred to as lower peristalsis = slower movement of stool through the colon. This is considered a neuromuscular problem; hence bowel movement will be slower overall. With this type, dietary strategies may not offer much help, especially when it comes to increasing your fibre intake.
- Outlet constipation or pelvic floor dysfunction: This is due to harm in the pelvic floor muscles supporting the bowel and bladder which causes the passing of stool through the bowel to be slower. Pelvic floor damage can happen due to pregnancy and child birth.

Secondary constipation can be the result of factors that affect bowel movement, including medications, neurological disorders (e.g. stroke, Parkinson's disease), diabetes, hypothyroidism, depression (due to the treatment) and an obstruction (in the case of cancer, for example).

Many different symptoms can make it complicated to find the right diagnosis and, due to the lack of standardised practices, it can be tricky for health professionals as well.

Occasional constipation is a fact of life. We all go through it. Most of the time, constipation is short-term but if you are experiencing a severe case of constipation coupled with one of the following symptoms, it is important to contact a healthcare provider. (Please do not self-diagnose or consult 'Dr Google' as you'll simply be wasting time and energy):

- Intense and/or abdominal pain
- Vomiting
- Persistent bloating
- Blood in your stool.

All of the symptoms above could mean there are larger health issues at play or that there is a benefit to getting immediate help from a health professional. Some serious conditions that may need to be ruled out are intestinal obstructions or perforations, or a large amount of hard stool may get stuck in your colon and can't be pushed out.

Establishing the type of constipation you suffer from is crucial for mapping out your action plan because what may help one type may worsen symptoms with another. We will get to that later.

Stool softeners and laxatives

Even though traditional and non-pharmacological treatments remain the mainstay for normal-transit and slow-transit constipation,

I want to equip you with some knowledge of common medications and agents used to treat chronic constipation:

Bulk-forming laxatives: These types of laxative refer to both soluble fibre and insoluble fibre, but the most commonly used are from soluble-fibre sources. The fibre absorbs a large amount of fluid within your colon, in turn, forming soft, bulky poo. The bulky stool causes your intestinal muscles to contract in order for the contents to move along and, *voila*, an easier bowel movement follows. Examples of these bulk-forming laxatives (i.e. types of fibre) include psyllium, inulin, methylcellulose, wheat dextrin and polycarbophil. As we've mentioned previously, it is essential that additional fibre is added gradually with enough fluid since common side effects of increasing fibre too quickly without the additional fluid include bloating, stomach cramps and even increased constipation.

Osmotic laxatives: These agents tend to be prescribed as an option for first-line treatment of constipation and they include polyethylene glycol, lactulose, sorbitol, magnesium hydroxide, magnesium citrate and sodium phosphate enemas. These laxatives draw water into the bowel from your intestinal lumen (i.e. nearby intestinal tissue) softening your stool and speeding up intestinal movement. Some of these laxatives can draw out nutrients such as minerals with the water and may cause electrolyte imbalances and dehydration.

Stool softeners: As the name suggests, these agents make it easier for water to be incorporated into your stool, making it soft and easier to evacuate. They are very popular after giving birth, post-surgery or throughout the joys of dealing with haemorrhoids, to avoid the need to strain. There is talk around the block though (i.e. the scientific kind) that stool softeners, for example docusate, are no better than placebo when it comes to relieving constipation, so nowadays, your doctor may look at more effective options.

Stimulant laxatives: I'm just going to start off by saying that these laxatives come with risky side effects and clinicians generally hesitate to prescribe these unless extreme cases warrant their use. Examples of stimulant laxatives include glycerine, castor oil, senna, bisacodyl and picosulfate. These stimulants act on your intestinal muscles to increase muscular contractions (peristalsis) moving contents along fast to reduce the transit time in your colon. These laxatives are not intended for long-term use as they can cause a serious glitch in how your colon functions, leading to laxative dependence for achieving peristalsis. Please use stimulant laxatives under the supervision of your clinician.

The con of senna

Did you ever come across 'skinny teas' or, as I'd like to call them, 'shit-yourself' tea blends? Senna has been a popular choice and dubbed 'nature's laxative'. It originated in Egypt and has been used for centuries as a laxative or gut stimulant. Nowadays, you'll see senna as an essential ingredient marketed as part of weight-loss teas and so-called detoxes.

Senna comes from the Cassia plant and the herbal mix is made from the leaves, flowers and fruit. Despite being approved as a remedy to treat constipation (if all other options, such as bulk-forming and osmotic laxatives, have failed), it comes with serious consequences and specific dose guidelines. Despite the warnings about abusing senna tea, it still seems to be trendy amongst young adults concerned about their weight and those suffering from eating disorders, where laxative abuse is a frequent compensatory behaviour. Long-term use of senna can cause dehydration and electrolyte imbalances (when levels of minerals such as sodium, potassium and magnesium get too high or too low), which in turn can lead to muscle spasms, weakness, irregular heartbeat and liver damage.

Senna should not be used for more than one week as another unpleasant side effect of long-term use is malfunctioning of the large intestine, causing a 'lazy bowel', just as I've noted above.

Enemas vs. colonics

You may also have heard of enemas and colonics? Just to ensure that I've covered every term you will come across if you suffer from constipation, let's get these two out of the way before diving into dietary and lifestyle approaches.

An enema is a procedure in which, in simple, graphic terms, a tube is inserted into your rectum via your anus, and liquid is injected (normally water) to stimulate your bowels to empty fully. It is generally used to empty your insides prior to surgery, if indicated, or for the occasional relief of severe constipation.

An enema may be one of the most uncomfortable feelings I've personally experienced and the reason I had to expose my bum to this form of flushing was to aid labour as my daughter made her journey earth side. An enema takes the term 'urgent' to another level, where in minutes, you feel like your insides are going to drop out and indeed they do within a few blinks. So yes, this dietitian has had her bum out on all fours, with a tube up there while casually chatting to a midwife about life.

A colonic, or colonic irrigation, differs from an enema in the amount of fluid used. In the case of an enema, we are looking at a one-off infusion of fluid whereas with a colonic, it is continuous – in other words, multiple infusions are administered for about 45 minutes.

Colonics are heavily marketed by the wellness unicorn industry and 'gut health' social media gurus claiming to cleanse your bowel, ridding it of toxins, which by now you should realise is utter nonsense and lacking in any scientific backing. While there is a space for colonics in some medical settings, unfortunately the procedure has been heavily commercialised towards those seeking a squeaky cleansed bum to feel light and rid of waste.

Colonics cannot be done at home since special equipment is required, including a trained professional. This glorified procedure comes with a ton of risks including electrolyte imbalances, bowel perforation, disturbing the balance of your gut microbiota, dehydration and infection to name a few. My suggestion is to let that part of your digestive system perform its duty as part of cleansing without external aid unless it is indicated by your health professional.

Enough rambling and into our dietary and lifestyle toolbox we shall dive! When it comes to managing constipation, Figure 12 represents a summary of the key changes you'll need to adopt.

Fibre

When it comes to fibre and constipation, if you've read Chapter 4 you'll be aware that your intake should be increased gradually to avoid symptoms such as gas, bloating and becoming even more clogged up. Follow my four-week strategy on page 81, but if you suffer from slow-transit constipation or pelvic floor dysfunction, this step may not offer much benefit and may make matters worse.

Slow-transit constipation is a neuromuscular problem so if things are not moving as they should due to a glitch in the nervous system, adding more fibre and bulk is just going to worsen the constipation. That doesn't mean you shouldn't be mindful of your fibre sources but, as I always say, it's important to know what type of constipation you're dealing with first to choose the right approach.

For normal-transit/functional constipation, here are some additional strategies you can try:

- Choose wholegrain over refined grain products. Choices include: oats, wholegrain pasta, barley, oats, brown rice, wild rice, quinoa and wholegrain bread varieties.
- Add ground flaxseed, chia seeds or bran to cereals, yoghurts and soups.
- Consume the skins of your fruit and vegetables when possible, such as those on potatoes, carrots, cucumbers, aubergine, apples, plums and peaches.
- Make your own trail mix with unsalted walnuts, pumpkin seeds, dried apricots and candied ginger.
- Add grated or mashed vegetables, such as carrots, courgettes and beetroot, to muffins.
- Add a three-bean mix to your salads and stews.
- Choose high-fibre snacks such as roasted chickpeas or steamed edamame.
- Experiment with legume-based dips such as hummus or white bean dip with crudités.

How much fibre?

It's useful to know what 30 grams (g) of fibre a day look like. Here is a guide:

- 40 g oats = 3 g fibre
- 200 g banana = 4 g
- 200 g orange = 4 g
- 50 g chickpeas = 5 g

- 2 wholegrain crisp breads = 4.5 g
- 60 g quinoa = 5 g
- ½ avocado = 6 g.

For slow-transit constipation and pelvic floor dysfunction, a good fibre target would be around 20-25 g to ensure that you're still getting some fibre but not so much that the poo-traffic builds up. Suitable food sources of fibre in these circumstances include:

- Having smooth nut butters instead of whole nuts.
- Include some legume-based dips instead of whole chickpeas and beans in a meal.
- Enjoy a smoothie both green and fruit based.
- Include more cooked vegetables than raw.

Toilet habits

My clients are always caught off guard when I ask them to demonstrate how they sit on the toilet. From now on, use the SEE method (Figure 12) for easy, breezy bowel movements (sit, elevate, eliminate). Sometimes you just have to SEE things a little differently to get better results.

When you sit on the toilet with your knees at a specific angle, your muscles relax partially, but when you are in a squatting position, your muscles relax completely. Elevating your legs simulates that squatting position, allowing you to have an easy bowel movement.

The other important habit to break is delaying the urge to go. Holding in your poo can be... well, uncomfortable to say the least. Generally speaking, holding it in on occasion won't cause you much harm; however, making it a habit to hold in your stool can cause unwanted health effects.

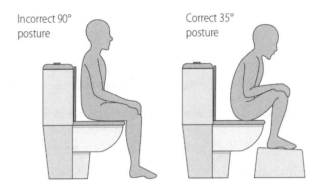

Figure 12: Sit, Elevate, Eliminate (SEE)

For example, holding in a stool for too long can cause your stool to turn hard, which makes going to the bathroom difficult and uncomfortable. For this reason, holding your poo for long periods of time can cause constipation, haemorrhoids and, in some instances, anal fissures. The other consequence of constantly delaying the urge to poo is something called rectal hyposensitivity. Nerve damage around the rectal area occurs and a glitch in how your colon and brain communicate causes you no longer to feel the urge to go.

Do you poo mindfully?

You've heard about mindful eating but what about mindful pooing? Here's the thing – so many of us 'do the deed' whilst being on our phones, scrolling through, and not even paying attention to our breathing or our posture. There are also many of us out there who are rushing and just want to get it over and done with, not fully emptying the tank. So here are some pointers to get you started with mindfulness on the loo:

1. No distractions! Be fully present and allow your body to relax, especially your pelvic floor muscles.

2. Engage in deep-relaxation techniques while defaecating.

3. Don't forget to SEE.

4. Avoid straining.

5. Do not stay on the toilet for more than 5–10 minutes.

Keep your phone at the door and lay paperbacks aside to ensure a good ol' mindful poo!

Kiwi fruit

Can a kiwi fruit a day keep constipation at bay? Well, two kiwi fruit actually can. There were a number of studies looking into managing constipation with the addition of kiwi fruit and they've all yielded positive results. Two kiwi fruit a day seems to have produced softer, bulkier poo, improved stool frequency and overall satisfaction with bowel habits. What makes kiwi fruit so special? Possible reasons include:

- Their high fibre content.
- Their high water-holding capacity, making stools bulkier and laxation easier.
- Their actinidin protease content; this is an enzyme with the role of protein digestion. Actinidin has been postulated to play a role in easing laxation.

If you aren't allergic to kiwi fruit, you can try adding two a day for four weeks and observe whether that makes a difference.

Fluid

Funny how we've always heard that you need to drink more water to manage constipation but did you know that this suggestion as a standalone is useless? Increasing your fluid intake makes complete sense if you are dehydrated and perhaps drinking less than 1.5 litres a day. Also, if you do plan to increase your fibre intake, you will definitely need to up your fluid intake due to fibre's action of absorbing water from your bowel. As a rule of thumb, aim to consume 1.5–2 litres of fluid daily, preferably water, noting that individualised requirements exist – that is, you may need to drink more depending on your activity levels, climate... etc.

Movement

Constipated or not, movement is a core pillar when it comes to managing any gut condition and for overall wellbeing. Just as many of us have a turbulent relationship with food, many also have a turbulent relationship with exercise. Try switching terminology to movement and ask yourself, why do you want to move?

Many of us have embarked on a training regimen or exercise plan for aesthetic reasons and weight loss. In my practice, this has been the least successful motivator to maintain a form of movement that's right for you, in the long term. In the case of constipation, we do know that inactivity is a major risk, so let's get moving.

Movement, whether a 20-minute walk or a spinning class, helps stimulate your gut muscles to contract, in turn, causing stool to move quicker. This way, poo moves faster through your large intestine rather than sitting there, absorbing so much water from your bowel as to cause hard and dry stools. The guidelines suggest we should aim for 150 minutes of moderate- to vigorous-intensity activity per week. My suggestions?

- Pick a form of movement that feels right, is enjoyable and that you're likely to keep up.
- Start by aiming for 15 minutes of movement daily. That can include a 15-minute walk, 15-minute spin session on a stationary bike or a 15-minute yoga flow.
- Forms of movement include walking, spinning, outdoor cycling, yoga, Pilates, swimming, rowing or dancing.

Do you need to see a pelvic floor specialist?

Clients suffering from chronic constipation are generally advised by yours truly to get an assessment by a pelvic floor physiotherapist. This is to ensure that the 'plumbing' down there is working and that we're not dealing with an underlying issue that is impacting the evacuation of poo. It also makes sense to work with a pelvic floor specialist given the role our pelvic floor muscles play when it comes to defaecation. I will go through some pelvic floor exercises and discuss the role of these muscles in detail in Chapter 12, but don't be surprised if your healthcare professional suggests you need some bum physio. In my practice, I regularly refer clients to a pelvic floor physiotherapist, which has made great improvements in my clients' bowel movements.

Supplements

There are numerous constipation aids out there but to keep things simple, I've highlighted the most popular choices to get unplugged and ones I frequently prescribe to clients.

Psyllium

Psyllium is a seed that comes from the plango plant, which grows mainly in India. Two teaspoons of psyllium provides us with 7 grams of fibre, so we are surely talking about a high-fibre seed. The main benefits of psyllium include:

- Helps you feel fuller for longer due to its high fibre content.

- Helps manage both constipation and diarrhoea.

- Can help lower cholesterol levels and protect against heart disease due to its high beta-glucan content, a type of fibre.

A total of 10-15 grams of psyllium per day is enough to exert these benefits and up to 30 grams per day is considered a safe amount (keeping in mind that too much fibre too quickly can cause gut discomfort). How to take it?

- My most common suggestion is to dissolve 1 teaspoon of psyllium husk in a glass of water, followed by another glass of water to make sure you've washed it all down. You can kick off your morning routine with this psyllium drink before breakfast.

- Mix it with your milk or yoghurt (or dairy-free alternatives).

- Enhance your porridge or smoothie with it.

- Sprinkle it on your salad.

- Use psyllium husk powder in baking (ideal to produce gluten-free baked goods).

Partially hydrolysed guar gum (PHGG)

Guar gum is a fibre extracted from the seeds of the guar bean. As the name suggests, guar gum is partially hydrolysed (broken down) using enzymes to lessen its thickening properties, improving its digestibility. PHGG is a water-soluble fibre, meaning that it dissolves in water to form a gel-like mass, and it has a low viscosity.

It has been suggested that this type of fibre has prebiotic properties and can be helpful for people experiencing constipation. Why is PHGG beneficial?

- It regulates bowel movement.
- It helps support the growth and activity of beneficial microbes.
- It can help regulate blood sugar and cholesterol levels.
- It has a satiating effect, making you feel full.

PHGG is different from guar gum in terms of its 'viscosity'. Guar gum is added to many commercial products generally as a thickener as it swells with water. This can be problematic for those with constipation or anyone at risk of intestinal blockages. PHGG, however, is much less viscous, minimising these side effects. How to take it?

- The dosage ranges between 3 and 8 grams per day, but do not exceed 12.
- In powder form, start with 3 grams dissolved in a glass of water and drink immediately, before a gel forms. Gradually increase your dosage to 5 grams per day and continue at this level for four weeks. If that dosage appears to resolve your symptoms of constipation, then you can continue at that dose for 12 weeks in total.
- If symptoms haven't fully resolved, you can try to increase the dosage to 8 grams per day, monitoring any symptoms of gas or bloating.

Probiotic supplements

As I've mentioned multiple times, choosing the right probiotic supplement comes down to the specific strain and the health concern we are attempting to resolve. The strains that have been well researched in aiding bowel frequency and symptoms of constipation include:

Bifidobacterium infantis 35624: At a dosage of one hundred-millionth (1 x 108) CFU taken for four to eight weeks.

Bifidobacterium animalis subsp. lactis, BB-12: Dubbed the world's most studied strain, BB-12 has been shown to improve stool frequency and help with constipation relief at a dosage ranging between 5 and 10 billion CFU for four to 12 weeks.

We've managed to go through steps to get unplugged but what if our problems are at the other end of the spectrum and we need to get corked? I'll see you at the next chapter that's all about diarrhoea.

Figure 13: Your constipation toolbox

8

The one on bumfluenza

My partner and I had been dating for about a year and here we were about to board a train from Alexandria to Cairo, with the built-up excitement of seeing the pyramids; a first for him and a third time for me. Wanting to make the most of our time, we organised a private tour of the pyramids with pick up awaiting us at the train station upon arrival. I remember hating myself for not packing any snacks so naturally, an hour into our trip, the 'hunger' pangs were hard to ignore. Being the brave soul that I was, I ordered a traditional cheese sandwich from the coffee cart roaming back and forth the isles. With every mouthful, I inched closer to a stomach apocalypse that was about to ensue.

Forty-five minutes out of Cairo, the sweats, cramps and rumbling commenced. Now, if you've been on these trains back in the day, you would avoid any toilet breaks like the plague. Unfortunately, there was absolutely no way of my retaining anything no matter how hard I clenched. I simply dragged Carlo, and sprinted towards the door labelled WC – that is, a hole in the floor on a moving train. To save you the details, I had one of the worst poo experiences of my life on a moving train with a man that I was dating manning the door and asking whether I needed support to 'balance' myself. Diarrhoea has no consideration of place and time so as we approached our final destination, I had to find a way to survive a day at the pyramids without soiling myself or dropping dead from dehydration. Our driver was lovely and so considerate, making frequent stops at every decent hotel en route to the pyramids to make sure I had access to 'modern toilets' as he so kindly put it. Rehydration involved the only thing I could get my hands on: salt, bottled water, Sprite and salty crisps. That's about all I was able to stomach throughout our tour of one of the greatest wonders of the world, in a horse chariot, with a horse suspiciously suffering from its own gut issues...

Defining diarrhoea and its causes

If you're a type 6 or type 7 on the Bristol Stool Chart (page 94), then we're dealing with loose poo or, as I like to call it, bumfluenza. In terms of its actual clinical definition, we are looking at two distinct types when it comes to diarrhoea:

- **Acute diarrhoea** is defined as having three or more loose or watery stools per day, which resolves within three to seven days. Diarrhoea is considered acute if it is of a duration of less than four weeks. This tends to be the most common form of diarrhoea.

- **Chronic diarrhoea** is having three or more loose or watery stools per day, which lasts for more than four weeks. If your diarrhoea lasts for more than 14 days, there is a chance that you may be dealing with something more than just an acute bout of loose poo. With chronic diarrhoea, we are looking at a range of under-lying issues that might be the cause.

The most common causes of acute diarrhoea are mainly infectious in nature, meaning you've possibly contracted a bacterial, viral or parasitic infection. We've all heard of the 'stomach flu', also known as viral gastroenteritis, and that tends to be one of the most common causes. Medication is another well known cause of acute diarrhoea and antibiotics are a familiar example. Most antibiotics, such as broad-spectrum antibiotics and erythromycins, are known as major disruptors of your gut microbiome given their mode of action, and this is one of the reasons why so many people taking them experience diarrhoea (i.e. antibiotic-associated diarrhoea). This can be prevented once we talk about using probiotics for treatment.

Chronic diarrhoea is quite complex in nature as there is a long list of conditions that need to be ruled out. As a dietitian, it is imperative that my clients with chronic diarrhoea get a diagnosis and this is where education and guidance can make a difference to gain control of your symptoms.

Loose poo investigations for chronic diarrhoea

What tests are generally recommended as an investigative starting point? When my clients approach me with chronic diarrhoea and if they haven't seen a specialist yet, this is the list of investigations I would recommend:

- Full blood count
- CRP (C-reactive protein, which is a marker of inflammation)
- Urea and electrolytes
- Coeliac screen
- Iron and ferritin levels

- Vitamin B12
- Faecal calprotectin (a marker of inflammation)
- Faecal culture and parasites
- Faecal elastase (to check pancreatic function).

Depending on the results of your investigations, a colonoscopy (a camera up your bum) may be warranted.

What are we trying to look for and rule out? Please keep in mind that, although some conditions can be scary when learning about what they are, more often than not, the cause of your chronic diarrhoea is manageable and not life-threatening. The following lists what may be the cause:

- Coeliac disease
- Inflammatory bowel disease (IBD)
- Irritable bowel syndrome (IBS)
- Small intestinal bacterial overgrowth (SIBO)
- Bile-acid malabsorption
- Food intolerances and allergies
- Infections such as *C. difficile*
- Pancreatitis
- Bowel cancers
- Side effects of medications:
  - Antibiotics, in particular macrolides (e.g. erythromycin)
  - Non-steroidal anti-inflammatory drugs (NSAIDs, e.g. ibuprofen)
  - Magnesium-containing products
  - Hypoglycaemic agents (e.g. metformin).

Food-related causes of chronic diarrhoea

Before we get into the detailed treatment of diarrhoea, I've listed the possible food related causes of chronic diarrhoea and a description of how diet is used to manage each. Some of the causes are discussed individually in the chapter such as IBS but here's a pocket guide you can use:

IBS and functional diarrhoea

These two common causes of diarrhoea generally respond well to dietary changes. The mechanism behind symptom development includes increased gut motility, visceral hypersensitivity and a disrupted gut microbiome. Dietary management will include a trial of the FODMAP process (i.e. reducing the amount of fermentable sugars in your diet – see page 142), improving eating habits such as small more frequent meals or even investigating the possibility of a food chemical intolerance, which can involve a low chemical diet.

Lactose intolerance

It is not uncommon for many people to cut out dairy when they experience any digestive problems and, as a matter of fact, as we get older we do start producing less of the enzyme called lactase that is required to digest lactose (milk sugar) in the small intestine. Symptoms of lactose intolerance include stomach pain, bloating and diarrhoea. Diagnosing lactose intolerance involves genetic or hydrogen breath testing with the avoidance of lactose-containing products followed by a challenge to confirm an intolerance. Dietary management of lactose intolerance does not necessarily mean complete avoidance of lactose, as some people are known to tolerate a small amount per day (up to 12 grams, which is equivalent to a 250 ml glass of milk). A low-lactose or lactose-free diet can be followed, ensuring that you do not miss out on important nutrients such as calcium.

Bile acid malabsorption (aka bile acid diarrhoea or BAM)

Bile acids are made in the liver, stored in the gallbladder and then released into the small intestine when food is eaten. Around 97% of these bile salts are recycled/reabsorbed and head back to the liver. BAM is when bile acids do not get reabsorbed in the small intestine and instead continue onto the colon, causing irritation and excess water secretions that lead to loose stools. BAM can be caused by a structural issue or by inflammation in that last part of your small intestine or can be secondary to another condition, such as SIBO, or after surgical removal of the gallbladder.

Symptoms include frequent diarrhoea and pale and greasy poo that is difficult to flush away, as well as stomach pain and bloating.

Generally, those diagnosed with BAM will be prescribed a type of medication called a bile acid sequestrant. You would also need to manage the fat content in your diet, to control your symptoms. Following a low-fat diet, consuming no more than 40-50 grams of total fat per day, can come with some risks such as the malabsorption of fat-soluble vitamins (vitamins A, D, E and K) as well as unintentional weight loss. For this reason, it is essential to work with a dietitian if you've been diagnosed with BAM to prevent this.

Coeliac disease

Coeliac disease is an autoimmune condition where there is a serious sensitivity to gluten, which is a protein found in wheat, rye, and barley. Eat gluten and your immune system goes on the attack and can damage your small intestine. Symptoms may include abdominal pain and bloating, diarrhoea, vomiting and weight loss. Other, more serious symptoms include anaemia, fatigue, bone loss and depression. What are the general guidelines for the treatment of coeliac disease?

1. Strictly eliminating gluten, which is found in cereal grains such as wheat, rye, barley and triticale. Alternatives include gluten-free cereals such as oats and rice and other gluten-free products that are high in fibre.

2. Ensuring you continue to follow a well-balanced diet that includes all major food groups within recommended guidelines. These are gluten-free breads and cereals, fruit and vegetables, meat and dairy, including their alternatives.

3. Label reading is a skill that you'll need to acquire. It is essential that you are able to identify food products that contain gluten, especially with regard to hidden ingredients, such as wheat-based thickeners and additives.

4. Avoid potential contamination of your food by using separate cooking utensils, appliances and cutting boards, as even minimal traces of gluten in your diet can delay gut recovery.

Small intestinal bacterial overgrowth (SIBO)

Did you know that the symptoms of IBS are similar to those experienced by people with SIBO?

SIBO is diagnosed when there are excessive numbers of bacteria within the small intestine. In a healthy digestive system, the number of gut bacteria within the small intestine is quite low, so when the number is excessive it is often the result of some anatomical abnormality or digestive illness. The most common causes are related to motility disorders (i.e. conditions that slow movement of food through the small intestine), structural problems present along the small intestine, such as adhesions, or medical conditions that can impact the motility of food in the small intestine, such as Crohn's disease and diabetes.

It is estimated that anywhere from 4% to 78% of patients with IBS also have SIBO. That's a pretty wide gap! But this is due to the fact that the medical community hasn't yet landed on one standardised test for diagnosing SIBO. Symptoms of the condition include bloating, abdominal pain or discomfort, flatulence, diarrhoea and/or constipation.

How is SIBO diagnosed?

Most cases of SIBO are diagnosed through the use of hydrogen breath testing. There are two types of these tests used: the lactulose breath test (LBT) and the glucose breath test (GBT). Both tests measure concentrations of hydrogen and methane in the breath.

If this is a test prescribed for you, you will be asked to drink a sugar solution and then provide samples of your breath at various intervals. If hydrogen and/or methane are detected within 90 minutes, SIBO will be diagnosed.

The length of time is key – as it generally would take two hours for the sugar solution to make its way to the large intestine, any rise in these gases prior to that time suggests that the sugar was acted upon by bacteria within the small intestine. This method has considerable diagnostic limitations: it lacks sensitivity and specificity, which means there is a high chance of getting a false positive or false negative result.

The best method of diagnosing SIBO is via sampling fluid from the small intestine and growing the bacteria found in this fluid to estimate if a very high number of bacteria are present. Because it is an invasive procedure, it is rarely used in clinical practice unless a gastroenterologist thinks it's necessary.

How is SIBO treated?

After you've been diagnosed with SIBO, the primary form of treatment is the use of a certain type of antibiotic; this is one that is not absorbed in the stomach and therefore can make its way to the small intestine where it can eliminate any bacteria it finds there. The most common one used is called Rifaximin aka Xifaxan.

Please note that SIBO can recur even after you have been successfully treated initially, so getting the right treatment plan is crucial.

Did you know you can manage symptoms of SIBO with oregano oil?

As a natural antibiotic, oregano oil is a great solution for bacterial infections, and specifically for those who suffer from SIBO or a variety of gut issues. How? Oregano oil's main active ingredients are believed to be thymol and carvacrol.

Carvacrol has the ability to act as an effective antibacterial that fights against a variety of bacterial strains while thymol acts as an anti-inflammatory agent in addition to its antibacterial, antiviral and antifungal effects. Due to the properties of oregano oil, it's a pretty good natural alternative for tackling the effects of SIBO.

This is the eradication plan for using oregano oil, but you will need to clarify with your gastroenterologist or dietitian that this approach will work for you:

- Use oregano oil capsules that provide 180-200 mg per capsule.
- Frequency: three times per day with meals.
- Duration: three to six weeks.

Because SIBO shares almost all the same symptoms as IBS, as you will see in the next chapter, researchers suspect a low-FODMAP diet may be beneficial for SIBO patients since this has had much success in IBS patients. A low-FODMAP diet would cause the problematic bacteria to 'starve' in the small intestine.

Therapeutic approaches to treating chronic diarrhoea

As this isn't a book about medication and treating your gut with drugs, I just want to offer a snapshot of some of the common pharmacological approaches to controlling diarrhoea when medication is required. These options are set out in Table 4. When it comes to the chronic type, every experienced specialist will agree that it is crucial we get to the bottom of what the real problem is and how the patient's quality of life can be improved by measures other than medical treatment.

Suggested dietary approaches to managing diarrhoea

Dietary management of diarrhoea will be dependent on the cause. Nevertheless, the first line of therapy will always be correcting any electrolyte and fluid losses.

Table 4: Common treatment options in chronic diarrhoea

Type of drug	How it works/ indication	Example(s) of agent	Dosage
Opiates	Inhibit gut motility and treat symptoms for most cases of chronic diarrhoea	Loperamide	2-4 mg, 4 times daily
Antibiotics	Treat conditions such as SIBO, *C. difficile*	Rifaximin, vancomycin, metronidazole	Individualised by your health practitioner
Bile-acid binders	Bind bile acids and bacterial toxins. Part of the treatment of BAM, ileal disease or diarrhoea associated with gall bladder removal (i.e. postcholecystectomy)	Cholestyramine Colesevelam	4 g, 1-4 times daily 625 mg, up to 6 times daily
Fibre supplements (not a drug)	Water-binding actions	Calcium polycarbophil Psyllium husk	5-10 g daily 10-20 g daily

Step 1. Rehydration

Fluid will be your best friend for the next 24 hours and the aim is to make sure you drink about 2-2.5 litres per day to make up for fluid losses:

- It is always best to make sure you have some oral rehydration solutions (ORS) in your home 'pharmacy cabinet'. ORS are commonly bought as sachets or tablets and contain a mix of glucose, sodium and potassium to prevent dehydration.

- Ensure you sip on ORS throughout the day.
- Other fluid choices include: bottled water, broth, diluted juice or sports drinks that contain salts such as sodium and potassium. It is best to avoid beverages such as sodas, fruit smoothies and milk-based beverages as they can worsen your diarrhoea.

How to make your own rehydration solution: A recipe from the World Health Organization

If you want to make your own rehydration solution, here's what the WHO recommends to make a 1-litre (l) solution using household measuring spoons for the quantities:

½ teaspoon table salt
½ teaspoon baking soda
2 tablespoons table sugar
1 litre of tap water, but only if safe to drink; otherwise, choose bottled water

How do I know if I need to drink more or if I am still dehydrated?

These are the signs of dehydration:
- The obvious – being thirsty
- Having dark-coloured pee
- Urinating less than normal
- Feeling extremely tired
- Dizziness.

1. Add 500 ml of water to your bottle/container.

2. Add the dry ingredients and stir or shake.

3. Add the remaining water to make your 1 litre home ORS.

Step 2. Eat easy-to-tolerate foods

If your diarrhoea and stomach cramps have settled, you're probably ready for some real food, but naturally you won't be rushing towards a creamy mushroom risotto or seafood chowder. You can start by including easy-to-tolerate foods that aren't too high in fibre; that doesn't mean, however, you should be eliminating fibre completely. As your diarrhoea eases, the goal is to temporarily reduce the amount of insoluble fibre you are consuming until your stool is better formed. That means you can still have soluble sources of fibre as listed below. We have moved on beyond the outdated approach of BRAT – bananas, rice, applesauce and toast – which lacks any scientific backing yet was widely recommended by paediatricians and doctors to temporarily ease diarrhoea.

Here's a basic food introductory guide as you manage diarrhoea beyond the BRAT recommendations:

Breads, grains and cereals

- Barley
- Gluten-free pasta
- Oats (cooked)
- Regular pasta
- Rice noodles
- Semolina
- Sourdough bread
- Spelt bread
- White bread
- White flour
- White rice

Vegetables (temporarily opt for well cooked vegetables instead of raw)

- Asparagus tips
- Carrots
- Courgette, zucchini
- Cucumber
- Mushrooms
- Potatoes (skinless)
- Pumpkin (skinless)
- Tomato paste and purée

Fruit (aiming for no more than two serves per day)

- Avocados
- Bananas
- Blueberries
- Honeydew melon
- Mango
- Oranges
- Pears
- Rockmelon
- Seedless watermelon
- Strawberries

Meat and meat alternatives (avoid tough, gristly meat)

- Chickpeas (¼ of a cup of canned chickpeas should be well tolerated)
- Eggs
- Lean beef
- Poultry
- Tofu

Dairy and dairy alternatives

- Temporarily choose lactose-free dairy options
- Natural Greek yoghurt
- Oat milk
- Soya milk and natural soya yoghurts

Nuts and seeds

- Smooth nut butter (e.g. almond butter)
- Chia seeds
- Ground flaxseeds.

Step 3. Limit foods that are known to exacerbate symptoms

Some foods are known to worsen diarrhoea because they can cause more gas and bloating as well as loose stool by causing food to move through your gut too quickly. While bumfluenza persists, it may be useful to avoid the following:

- Caffeine in coffee, strongly brewed black tea and chai, cola and energy drinks.
- Foods high in insoluble fibre, such as wholegrain or multigrain breads and cereals, wholegrain pasta, whole nuts, broccoli, Brussels sprouts, black beans, celery and lentils.
- Consuming high amounts of fructose, which is the sugar found in honey, dates, dried fruit and apple juice, and too much fruit in one go.
- Fried or fatty food such as anything fried in butter, French fries, potato crisps and pastries.
- Sugar alcohols found in diet foods, such as sugar-free gum, sweets and high-protein dairy products.

Step 4. Supplement wisely

When it comes to easing diarrhoea with supplements, we can certainly look at psyllium husk and a few probiotic strains. Here's how:

Psyllium husk: Yes, we've used psyllium husk for constipation but you can also use it to manage diarrhoea, thanks to its water-holding capacity in your bowel that bulks up watery, loose stools. This is the protocol I recommend:

- Day 1: Start with ½ teaspoon in a glass of water (250 ml), and then wash that down with another glass.

- Day 2: Increase to 1 teaspoon in a glass of water and wash that down with another glass.

Probiotic supplement choices

Lactobacillus rhamnosus GG (LGG). This probiotic strain is used to treat and prevent antibiotic-associated diarrhoea and traveller's diarrhoea. You can take this strain two hours after breakfast and dinner at 5-6 billion CFU, twice a day for the duration of your antibiotic course and for one week after.

Saccharomyces boulardii CNCM I-745. This probiotic yeast has been also used to prevent antibiotic-associated diarrhoea and traveller's diarrhoea. It is commonly sold as sachets at 10 billion CFU.

Enterococcus faecum SF 68. This is actually a strain commonly found in Europe that I have used with my clients for the prevention and treatment of any infectious type of diarrhoea. It generally comes in capsule form and the protocol for its use is:

- For diarrhoea prevention: Take 2 capsules daily for the duration of two weeks.

- For the treatment of diarrhoea: Take 3 capsules daily for the first week, then reduce to 2 capsules by the second week.

We've covered constipation and diarrhoea and now it's finally time to learn more about a condition that causes either or both. Irritable bowel syndrome, here we go!

9

The one on irritable bowel syndrome (IBS)

My story with IBS is not uncommon. In 2015, I struggled with horrendous gut symptoms ranging from diarrhoea and stomach pain to bloating and tummy cramps. I also suffered from anxiety. So I HAVE been there. I have been told that it's all in my head and I've had my social life hugely impacted. If you are reading this as a sufferer of IBS, I am here to tell you that:

YOU CAN beat the uncomfortable bloat.

YOU CAN find relief and enjoy eating out without fear.

YOU CAN live a life no longer controlled by your symptoms.

YOU CAN know your exact triggers.

YOU CAN restore your gut health.

An easy-to-digest start to understanding IBS

In my practice, IBS is one of the most frequently reported diagnoses that clients come to me with. If you are new to the topic, let me break it down into little nibbles that are easy to digest:

- IBS is a chronic condition that is more prevalent than we think, and of which the exact cause seems to be a combination of different things, which I will illustrate a few lines down.

- Prevalence is around 10-15% of the world's population.

- With IBS, there seems to be a glitch in how the gut-brain axis works, hence its being considered a functional gut disorder aka a disorder of the gut-brain interaction. This causes a disturbance in how your digestive system functions. Structurally, there is nothing wrong.

- The cardinal symptoms of IBS include abdominal pain, a noticeable change in bowel movements (diarrhoea, constipation or alternating between the two) and bloating.

- Diagnosis starts with a 'process of elimination' as conditions such as inflammatory bowel disease (IBD) and coeliac disease need to be ruled out. However, what are called the 'Rome criteria' are a useful tool for diagnosing IBS. Once a diagnosis has been made, then a number of approaches are considered to manage symptoms.

- Dietary triggers of IBS include FODMAPs, fatty foods, alcohol, caffeine and highly spiced foods. Non-dietary triggers include stress, anxiety and poor sleep, to name a few.

- Management is a multifaceted approach because the condition affects every aspect of your life. This is why I've created the four-pillar approach to managing IBS (see page 200), which includes addressing nutrition, mind, movement and sleep.

- The FODMAP process is a clinically supported approach to managing IBS symptoms; however, it is not a cure. It is also only a temporary diet and should not be followed in the long term.

Good communication and empathy go a long way: a note for healthcare practitioners

This sidenote is dedicated to health practitioners working with IBS patients and who happen to be reading this book. When it comes to IBS, it is evident that there is a mismatch between the views of patients and those of doctors when it comes to IBS, negatively impacting the patient-clinician relationship.

I have personally been told that it's 'all in my head', which has been a common statement delivered to my clients too. In one study, the majority of GPs reported that IBS was primarily a psychological problem. Doctors also appeared to have conflicting views on IBS – the publicly expressed 'medical definition' vs. their private view of IBS based on their own experiences of treating patients with the condition, plus a considerable helping of prejudice. So, what can doctors do? A recent review highlighted the following practices to adopt a better, more meaningful and engaged consultation with patients:

- Listen with intent and complete attention.
- Investigate what your patient cares about the most, setting priorities together.
- Seek to connect with your patient's story and acknowledge their actions and any successes.
- Explore emotions without dismissing emotional cues.

A common feeling amongst people diagnosed with a functional disorder, myself included, is that they feel stigmatised by a diagnostic label that is perceived as being less important than that of an organic disease. The symptoms are identical, the suffering similar and genuine, yet the support provided is less than adequate. Empathy and communication go a very long way. A positive and empathetic attitude has the potential to yield more positive outcomes for IBS patients.

The perceived causes of IBS

Despite being very prevalent, the exact cause of IBS remains unknown. Research is uncovering a number of culprits but it's likely a combination of causes. These are:

A glitch in the gut-brain axis: IBS is actually known for that glitch in how our brain and gut communicate causing issues with how gut muscles contract as well as how 'sensitive' our guts become. So, the communication line is somehow disrupted.

Changes in the gut microbiome: This is where I'm bringing the term 'dysbiosis' back, otherwise known as poor microbial diversity. As we saw earlier (page 19), many factors can affect your gut microbiome, such as the environment, diet, stress and medication, negatively impacting your inner ecosystem.

Intolerance of fermentable sugars: Many IBS sufferers struggle with digesting fermentable sugars (aka FODMAPs) found in things such as bread, onions, asparagus, dairy products and apples. This struggle can cause an IBS-impacted gut to become hypersensitive to the effects these foods can have.

Stress: We know that stress negatively affects the gut-brain axis but we also see the effects of stress on gut motility, our immune system, intestinal permeability and the diversity of our gut microbiome.

Post–infection: More common than we think, but post-infectious IBS is a thing. So many clients have reported how their guts have become the 'weakest link' after picking up a stomach bug while traveling. An infection can cause changes to your gut microbes and influence intestinal permeability.

As you can see, it's more complicated than one might think, hence you need a good support team by your side and that will include a gastroenterologist, a GP and a gut health dietitian (and possibly this book).

What are the types of IBS?

IBS comes in multiple forms: IBS-C, IBS-D and IBS-M/IBS-A (and post-infectious IBS as mentioned above).

IBS-C

IBS with constipation, or IBS-C, is one of the most common types of IBS. With this type, you'll experience fewer bowel movements overall, and you may sometimes strain to go when you do have them. We're also looking at a type 1 or type 2 stool on the Bristol Stool Chart (page 94).

IBS-D

IBS with diarrhoea, aka IBS-D, causes the opposite issues to IBS-C. If you have IBS-D, you may also feel abdominal pain along with more frequent urges to go. Excessive gas is also common. With this type, you'll be seeing more type 6 and type 7-like stools (page 95).

IBS-M or IBS-A

Some people have another type called IBS with mixed bowel habits, or IBS-M. IBS-M is also sometimes called IBS with alternating constipation and diarrhoea (IBS-A). If you have this form of IBS, your stools on abnormal bowel movement days will be both hard and watery.

Post-infectious IBS

Post-infectious (PI) IBS refers to symptoms that occur after you've had a gut infection. After your infection, you may still have chronic inflammation along with issues with gut flora and intestinal permeability.

I'm IBS-D, pleased to meet you...

How is IBS diagnosed?

Diagnosing IBS is not as straightforward as you might think given that there are no biochemical or physiological abnormalities. For this reason, the diagnosis is symptom-based and the most widely used diagnostic tool, as mentioned earlier, is based on the Rome IV criteria:

> *Recurrent abdominal pain on average at least one day a week in the last three months associated with two or more of the following:*
>
> * *related to defaecation*
>
> * *associated with a change in a frequency of stool*
>
> * *associated with a change in form (consistency) of stool.*
>
> *Symptoms must have started at least six months ago.*

Getting a diagnosis of IBS will also involve a typical work up, which looks like this:

- FBC (full blood count)
- ESR (erythrocyte sedimentation rate)
- CRP (C-reactive protein)
- Coeliac screen, which includes endomysial antibodies (EMA) or tissue transglutaminase (TTG)
- Stool bacteriology/parasite screen
- Faecal calprotectin
- Haemoccult (hidden blood in stool)
- Further interventions if indicated: abdominal ultrasound, gastroscopy/colonoscopy.

The alarm signs

Clients always ask the question: 'When should I be worried?' When it comes to symptoms that need to be addressed urgently and are highly unlikely to be indicative of IBS, keep these features/changes in mind:

- If you're 60 years or older experiencing a change in bowel habits to looser and/or more frequent stools persisting for more than six weeks.
- Unintentional weight loss.
- Blood in your poo aka rectal bleeding.
- Raised inflammatory markers for inflammatory bowel disease.
- Symptoms suggesting ovarian cancer (abdominal bloating, weight loss, pain in the pelvic region, changes in bowel movements...)
- Family history of bowel or ovarian cancer.

With a list of symptoms that are far from sexy, is there no sex with IBS?

Intimacy and IBS is a topic not talked about enough and in a survey of over 300 healthcare professionals, 79% agreed that IBS can affect a patient's romantic relationships.

The first thing to consider is that the symptoms of IBS are the furthest thing from sexy that you can get. At the beginning of any relationship, tummy and poo troubles are hard and embarrassing to talk about. Imagine having to rush to the loo when you've just arrived at your partner's place for the first time? Or how about showing any sort of intimacy when you're extremely bloated and are terrified of letting some gas 'slip-away'?

Another thing that people don't talk about is the effect of IBS on sex. Even if your sex life isn't directly affected by IBS, feeling confident and alluring is oftentimes difficult when you're bloated, in pain and thinking about where the closest bathroom is. Now, a lot of this can be addressed with honest conversation and an understanding, supportive partner. So, I encourage you to BE OPEN AND COMMUNICATE. Start a conversation about what you're going through and you may be surprised at the amount of support your partner is ready to give. I know I am grateful for mine as he has lived through it all.

Management of IBS

While there is no 'cure', IBS is a condition that requires a multifaceted approach in order to gain control of symptoms and not the other way around. Being diagnosed with IBS doesn't mean the end of social eating and losing control of your bowels and bloat. It may feel that way for a while, but once you've figured out the missing pieces in your gut puzzle, through education, this book and the right support, you will find relief and enjoy eating again sans fear.

As always, I only want to offer a snapshot of medical approaches to managing gut conditions (see Table 5), but with IBS, dietary and lifestyle modifications are generally the first line of therapy everyone is advised to undertake. If symptoms fail to improve after such changes, then a number of first- and second-line drug treatments are carefully considered.

Table 5: Medical treatments for IBS

First-line drug treatments	Second-line drug treatments
Antispasmodics (medication that relaxes troublesome gut muscle contractions) Peppermint oil Anti-diarrhoeals (e.g. loperamide) Laxatives	For persistent symptoms if dietary and lifestyle changes have failed to elicit any change as well as first-line treatments Tricyclic antidepressants (TCAs) Selective serotonin reuptake inhibitors (SSRIs) Second-line drugs targeting abnormalities in bowel habit

Why use TCAs and SSRIs? In simple terms, to target the gut-brain axis. Our central nervous system has the ability to alter the function of our gut, including motility and how sensitive we are to pain and bloating. We also know that anxiety and depression are common amongst IBS sufferers and the converse is also true; those with depression and anxiety are more likely to develop IBS. For this reason, such drugs, only when indicated, act on this bidirectional pathway, subsequently improving symptoms of IBS.

Peppermint oil: A global gut soother

An ancient remedy for many ailments, peppermint oil has been historically used for easing a turbulent tummy due to its antispasmodic and antiemetic effects. In layman's terms, it helps with stomach cramps and nausea. With regard to IBS, it has been used as an antispasmodic as it acts on the smooth muscles of the intestine. The active ingredient supplying tranquillity to your gut is called l-menthol. It is known to reduce muscle spasms in your gut as well as preventing the formation of gas and easing its expulsion or, as we'd like to say, helps you trump, toot and fart with ease. Peppermint oil is commonly sold as capsules and here's my guide to using it as part of managing your IBS:

- IBS subtypes likely to benefit: IBS-D and IBS-M.
- Dosage: 1200 mg daily – that is, 180-400 mg three times per day taken preferably one hour before meals.
- Duration: Four to 12 weeks.
- Contraindications: GORD (page 64), active stomach ulcers, pregnancy and lactation at high doses.

First-line dietary and lifestyle approaches

You're reading this book for diet and lifestyle advice, so let's talk about first-line strategies before embarking on any specialised approaches and eliminations. These are summarised in Table 6.

Table 6: Diet and lifestyle interventions for IBS

Type of intervention	Guiding points
Take care with fibre	We've raved about the importance of fibre and how we should be aiming for the recommended target of 25-30 grams per day but that, my friends, may not necessarily apply to you! Depending on the type of IBS, the 'how much' and 'type' should be adjusted according to your symptoms. There's a general consensus that insoluble fibre should be limited and that benefits seem to point towards consuming soluble fibre. Avoid supplementing with wheat bran too as that may worsen your IBS symptoms. For IBS-D: • Supplement with soluble fibre (e.g. psyllium husk (page 109)) • Cut down on insoluble fibre such as wholegrain breads and cereals and wheat bran • Find your fibre balance For IBS-C: • Supplement with soluble fibre (e.g. psyllium husk or partially hydrolysed guar gum (page 110)) • Gradually increase fibre intake from diverse sources • Add ground linseeds (flaxseeds) daily and increase gradually. The dosage is between 6 and 24 grams per day, so find that sweet spot steadily

Table 6: Diet and lifestyle interventions of IBS – Cont'd

Type of intervention	Guiding points
Avoid/ reduce alcohol	If you don't drink any alcohol, you can skip this part and there's no reason for you to start. If you do consume alcohol, here's how it can affect your symptoms: • It affects how your gut muscles contract and this is why many complain of diarrhoea after a big bender or even just a glass or two • It can weaken your intestinal lining, making your gut more 'permeable' i.e. leaky. This can impact the absorption of different nutrients • Chronic consumption can negatively impact your gut microbiota Combining the effects above is a perfect brew for worsening symptoms. If you do consume alcohol and you notice that it is a trigger, here's what you can do: • Swap cocktails for mocktails for 2-4 weeks • Reintroduce small amounts twice a week i.e. ½-1 glass at a time. That's a maximum of 1-2 glasses a week Do not overdo it and limit to no more than four standard drinks per week. What is a standard drink or 1 unit? 30 ml of a spirit, 125 ml of wine, 330 ml of beer

Limit fatty foods	People with IBS, especially the D type, notice a heightened sensitivity to fatty foods. This is mainly due to how fat can over-stimulate the gastrocolic reflex, which is your body's response to consuming food. Your digestive tract receives signals for your intestinal muscles to contract and move food along, making room for more. With IBS, due to the gut being overly sensitive, the contractions increase in intensity and fatty foods can be a culprit, causing stomach pain, diarrhoea and bloating. Limiting your consumption of fatty food can potentially ease your symptoms
Limit caffeine	This is a tricky one because we have no conclusive recommendations regarding caffeine intake and managing IBS symptoms. Caffeine is believed to increase gut muscle contractions in the colon and, more intensely, increases the activity of muscles around the rectal area. This is why many claim that coffee stimulates an immediate poo! Coffee consumption is related to symptoms such as GORD, indigestion, stomach pain and loose stools. If you notice that caffeine is a trigger, in simple terms, cut down. Limit your intake to one caffeine-containing beverage per day, preferably with or after a meal
Try specific probiotics	Probiotics have been widely used in people with IBS and are generally regarded as safe but it all comes down to the type of strain used. Certain strains will offer specific benefits depending on the type of IBS and the symptoms you're aiming to manage The following strains have been extensively studied: *Lactobacillus plantarum* 299v • Helps with bloating, gas, tummy pain, bowel regularity *Bifidobacterium lactis* BB-12 • Helps with straining, constipation, bowel regularity *Bifidobacterium lactis* HN019 • Helps with tummy pain, bloating, gas, loose stools, bowel irregularity

Table 6: Diet and lifestyle interventions of IBS – Cont'd

Type of inter-vention	Guiding points
Move more	There is no doubt that movement plays a positive and crucial role when it comes to our physical and mental wellbeing. When it comes to regular movement and our gut, being active: • helps prevent constipation • helps you beat the bloat by easily tooting gas • positively impacts microbial diversity • improves IBS symptoms by targeting the gut-brain axis • Power walking or light-jogging Pick your choice of exercise wisely though. Some forms of movement, such as HIIT (high intensity interval training) or competitive running and cycling, can make you feel worse, so perhaps avoid these temporarily and make the swap to: • Gentle spinning • Swimming • Yoga, Pilates or barre classes • Low-intensity strength training • Walking

The FODMAP process

We've briefly introduced you to the low-FODMAP diet when we spoke about bloating but I'm dedicating the next few pages to the FODMAP process. Now, I prefer to call the low-FODMAP diet the 'FODMAP process' because calling it a low-FODMAP diet implies that it is simply a long-term elimination diet when we know it includes three distinct phases.

To refresh your memory, the term FODMAPs stands for 'fermentable oligo-, di-, mono-saccharides and polyols'. These are fermentable carbohydrates that are resistant to human digestion.

Instead of their being absorbed into your bloodstream, your gut bacteria use these carbs for fuel, producing hydrogen gas and causing digestive symptoms in sensitive individuals. FODMAPs draw liquid into your intestine, which may cause stomach pain, bloating and diarrhoea, constipation or a mix of both.

The low-FODMAP diet is the most widely used elimination diet when it comes to controlling symptoms of IBS and is the primary dietary intervention for IBS sufferers who do not respond to the first-line dietary changes mentioned above. The three phases of the FODMAP process include:

Phase 1: Elimination phase

Phase 2: Reintroduction or challenge phase

Phase 3: Liberalisation or Personalisation phase

In more detail:

- In phase 1, you follow the low-FODMAP diet, which is a test diet eliminating high-FODMAP foods for two to six weeks to achieve symptom control.

- Phase 2 is all about the reintroduction of FODMAPs via challenges: different tester foods are tried to help you determine your threshold of tolerance and which FODMAPs you are sensitive to.

- Phase 3 is all about rejoicing and striking a balance between the FODMAPs you tolerated with no issues while limiting the high-FODMAP foods you are sensitive to.

The FODMAP process is now widely available all over the internet; an online search can pull out over a thousand results that offer lists and tables of foods to avoid and foods to include. You can also go through a self-guided elimination diet using an app. My advice though is to get some form of support from a qualified and experienced dietitian or registered nutritionist as there are pitfalls, which I will go through in detail, but first, let's meet the FODMAPs.

What are FODMAPs?

Fructose

Fructose is a simple sugar found mostly in fruit and some vegetables. Fructose malabsorption happens when your body cannot absorb fructose properly in the small intestine. Funnily enough, fructose can only be absorbed with the help of glucose. So, foods with a higher amount of fructose than glucose can be troublesome for many. So where is fructose found, exactly? In fruit such as apples, raspberries pears, peaches, mangoes and watermelon, and in dried fruit and fruit juices as well as sweeteners such as agave and honey.

Lactose

Lactose is the natural sugar found in milk and dairy products, such as:

- Milk
- Evaporated milk
- Yoghurt
- Ice cream
- Custard
- Certain soft cheeses such as ricotta, cottage cheese and mascarpone.

Fun fact: Lactose tolerance is a 'genetic mutation' and lactose intolerance is considered a normal occurrence. As we get older, we produce less lactase, which, as you now know, is the enzyme required to break down lactose. Thankfully, there is a plethora of lactose-free products available as well as lactase enzymes to help you digest lactose. (See section on lactose intolerance, page 117.)

Fructans

Fructans tend to be very troublesome for many people. And this is where 'I'm intolerant to gluten' can be thrown out the window. Fructans are oligosaccharides and polysaccharides that are found in a variety of vegetables and... drum roll... WHEAT!

Fructans are found in many foods, including:

- Vegetables such as asparagus, broccoli, beetroot, cabbage, chicory, garlic, leeks, onions and garlic.
- Grains such as wheat and rye.
- Added fibre, such as inulin and fructo-oligosaccharides, commonly added to items such as probiotic supplements, granola bars, and frozen desserts.

People who have trouble digesting fructans may experience gas, bloating, belching and constipation or diarrhoea.

Clients who come to me blaming gluten tend to have a fructan sensitivity instead. By saying that though, my goal is to identify whether this group is troublesome and, if so, establishing a threshold of tolerance – that is, how much can you tolerate before experiencing symptoms? It is also important to reintroduce this group down the line as it is a source of prebiotics for our gut microbes.

Galactans

Galactans are a type of fermentable sugars found in legumes (i.e. pulses): think baked beans, kidney beans, chickpeas and soya products. Many people fear this food group due to the tummy chaos that follows after a few mouthfuls. Galactans are malabsorbed because the intestine does not have the enzyme needed to break them down. Consequently, galactans can contribute to gas and gut distress. By saying that though, there is a way to introduce this group in small amounts to expose your gut to the beneficial fibre found in legumes. A digestive enzyme called 'alpha-galactosidase' that can be taken as a supplement can also help with digestion and minimising symptoms associated with beans.

Polyols

We made the acquaintance of polyols when we addressed artificial sweeteners and diarrhoea (page 82). Polyols such as xylitol, sorbitol, maltitol and mannitol are found in some fruit and vegetables and are often used as sweeteners in chewing gum, protein powders and 'diet'/sugar-free foods. Similar to fructose, polyols attract water as they move through the small intestine; this occurs whether polyols are absorbed or not, but it can lead to motility problems (hello, diarrhoea) for people who are more sensitive to the pressure this fluid exerts on the intestinal walls. The gas produced as a by-product of bacterial fermentation causes additional pain, bloating and altered bowel habits.

A FODMAP warning

Before jumping on the LOFO (i.e. 'low-FODMAP') bandwagon to address your IBS symptoms, keep in mind that it does have its fair share of cons:

- Compliance and limitation: It is a diet that is difficult to follow and can be challenging since it requires a lot of planning. The good thing is that it is only meant to be followed for two to six weeks (and not more).
- Nutritional inadequacy: Because you are eliminating a lot of different foods, you will be at risk of missing out on

important nutrients. It is therefore important to stick to this diet for only a short period of time.

- Inappropriate use: The low FODMAP diet is often misunderstood. It is not a long-term diet and it is not designed for weight loss, it is not necessarily healthy and should not be used without having consulted your doctor/ nutritionist/dietitian.

- Altered gut microbiota: The FODMAP diet cuts out important prebiotics, in turn exposing your gut flora to the possibility of less diversity. This can change the composition of your inner ecosystem, your gut microbiota, as the growth of 'good' bacteria can be reduced.

- Disordered eating: Because the diet is so restrictive it can also foster disordered eating or a restrained, unhappy relationship with food.

Where to start

Step 1. Food and symptom diary

Complete the food and symptom diary that is part of this book for one week.

Step 2. Looking for triggers and patterns

Ask yourself:

- Am I seeing any patterns?
- Is fibre an issue?
- How about caffeine?
- Is it just lactose that's triggering my symptoms?

By identifying patterns, you may be able to stick to first-line approaches without the need for a strict elimination diet. I would suggest trialling the first-line approaches for two weeks and, if you do not experience any improvement in symptoms, it may be worth moving on to step 3, which is the low FODMAP diet.

Important note: Before commencing, please note that this is only a simplified version of the diet and I strongly recommend you seek guidance from an experienced dietitian to ensure that you consume a diverse range of foods.

Step 3. The low-FODMAP diet (Phase 1)

Clients generally notice a difference within two weeks of following the elimination phase, as shown in Table 7. If you do not notice a difference after two weeks, then I would recommend you stop and reintroduce foods back in. The low-FODMAP diet has been shown to reduce symptoms in up to 75% of IBS sufferers, so there are a minority (at least 25%) of non-responders and that shouldn't surprise you. We are slowly uncovering the facts around the relationship between how well or poorly you respond to the diet and your gut microbes. So here goes...

Table 7: Foods to allow and to avoid on the low-FODMAP diet

Fruit	
Low-FODMAP – Allowed	**High-FODMAP – Temporarily avoid**
Bananas	All dried fruit, dates, goji berries
Berries – all types except blackberries	All fruit juices
Cantaloupe (no more than 150 g)	All stone fruit: apricots, cherries, peaches,
Grapefruit, lemons, limes	nectarines, plums
Grapes (limit to 15 at a time)	Apples
Honeydew melon	Blackberries
Kiwi fruit	Custard apple
Oranges, mandarins, tangelos,	Lycheese
clementines	Mango
Passion fruit	Pears, Nashi pears
Pawpaw/papaya	Persimmon
Pineapple	Watermelon
(fresh; no more than 150 g)	
Rhubarb	
Vegetables	
Low-FODMAP – Allowed	**High-FODMAP – Temporarily avoid**
Alfalfa sprouts, bok choy, bamboo	Artichoke
shoots, bean sprouts	Avocado, asparagus, beetroot
Aubergine, endive, ginger, green	Broccoli, Brussels sprouts,
beans	Cabbage, cauliflower
Carrots, corn, cucumber, chives, chili	Fennel
Choy sum, collard greens	Garlic
Lettuce (all types), olives	Leeks
Pumpkin (butternut), potato, pars-	Mushrooms
nips, radish	Okra
Spinach, silver beet, squash, seaweed	Onions, spring onions (bulb), onion powder,
Tomatoes, turnips, taro, courgette	shallots
(zucchini)	Sugar snap peas, snow peas, garden peas

Table 7: Foods to allow and to avoid on the low-FODMAP diet – Cont'd

Dairy and alternative products	
Low-FODMAP – Allowed	**High-FODMAP – Avoid**
Brie, camembert, blue cheese	Buttermilk
Lactose-free milk	Cottage and ricotta cheeses
Lactose-free yoghurt	Cow's milk (full and low fat)
Mozzarella	Cream and sour cream
Rice, oat and almond-based	Custard, ice-cream
alternatives	Kefir
Soya milk (if made from soya beans)	Mascarpone and cream cheeses
Yellow cheeses	
Meat, fish and eggs	
All types allowed (but avoid crumbed/floured ones or ones that use cornflakes or rice crumbs) Note: Many seasonings will contain onion powder; check ingredients.	
Breads and cereals	
Low-FODMAP – Allowed	**High-FODMAP – Temporarily avoid**
100% oat bread	Barley
Carman's fruit-free muesli	Burghul (cracked wheat)
Flours: millet, potato, rice, oat,	Couscous
buckwheat	Fruit breads
Most gluten-free products	Pastry
Pastas: rice, corn or quinoa	Rye breads and rye crackers
Rice (white or brown), rice noodles,	Spelt flakes
polenta	Wheat based pasta
Rice Mountain bread wraps	Wheat-based products such as breads
Rice or corn flakes/puffs	wheat crackers, biscuits, pastries, cakes and
Rice/corn cakes and crackers	desserts
Rolled oats/porridge, oat bran	
Spelt sourdough bread	

Legumes, nuts and seeds	
Low FODMAP – Allowed	**High FODMAP – Temporarily avoid**
Miso	Almonds and hazelnuts (can limit to 10
Nuts (Brazil nuts, chestnuts, maca-	nuts)
damia, peanuts, pecans, pine nuts,	Cashews
walnuts)	Lentils and legumes: chickpeas, kidney
Peanut butter	beans, baked beans (can limit to ¼ cup and
Seeds (chia, poppy, pumpkin, ses-	canned legumes are better tolerated)
ame, sunflower)	Pistachio
Tempeh (plain)	
Tofu (plain)	
Miscellaneous	
Low FODMAP – Allowed	**High FODMAP – Temporarily avoid**
Berry jams (NOT blackberry)	Apricot jams and apricot yoghurts
Chicken and beef stock cubes (that	Dessert wine, port
are free from lactose, gluten, onion	Doughnuts, pastries
and garlic)	Honey
Coconut cream and milk	Most commercially-baked items
Dark chocolate	Most flavoured potato chips/crisps and
Golden or maple syrup	rice crackers (many flavours contain onion
Marmalade	powder)
Mayonnaise	Most packaged/tinned soups (contain
Oils (including garlic infused oil)	onion)
Soy sauce, oyster sauce, chili sauce	Protein powders (many contain lactose)

Additional tips for following the low-FODMAP diet

All the following suggestions will make it easier and more pleasant:

1. Flavour, flavour, flavour!

   - Use spices or herbs, like basil, chili, cilantro, cinnamon, cumin, ginger, pepper, rosemary, tarragon or thyme, on meat, fish, chicken or vegetables.

- Use maple syrup instead of honey when baking or adding to cereals.

- Use chives, the *green part* of a spring onion or garlic-infused *oil* to amplify the flavours of your dishes when cooking.

2. Even when following the low-FODMAP diet, I encourage my clients to aim for 30 plant ingredients per week. Use the guide mentioned on page 88 and make sure you include low-FOD-MAP ingredients.

Basic low-FODMAP ideas to help get you started

These are my tried-and-tested suggestions for getting you started:

Breakfast:

- Corn- or quinoa-based cereal with lactose-free milk and strawberry slices
- Porridge with almond milk and blueberries
- Gluten-free toast with peanut butter or feta

Lunch:

- Green hearty salad: Kale, baby spinach, grated carrot, sliced cucumbers, 1 tablespoon of sunflower seeds and 2 sliced hard-boiled eggs.

- Quinoa salad: 3/4 cup of cooked quinoa with baby spinach, handful of walnuts, 6 olives, sliced cucumbers, parsley, cherry tomatoes, sliced carrot, 4 sliced radishes, 50 g of crumbed lactose-free feta and ¼ cup of canned (drained) chickpeas. Dressing: Lemon vinaigrette

- Chicken salad: 120 g of grilled chicken breast (shredded) with baby spinach, lettuce, cucumbers, green beans, cherry tomatoes, 1 tablespoon of pumpkin seeds, topped with a sweet chili dressing.

Dinner:

- Chicken or beef stew with carrots, courgette and spinach
- Tofu and ginger stir-fry with bok choy, courgette, spinach and aubergine
- Steamed fish with ½ cup of cooked brown rice and green beans and a side salad.

Phase 2: Reintroduction or challenge phase

If you have completed phase 1 successfully, with minimal symptoms for five days straight (as early as two weeks into the diet), the challenges can commence! There are different protocols out there when it comes to reintroducing FODMAPs; below I have simplified the approach as much as I can so as not to lose you.

Here are some basic guidelines to keep in mind:

- Yes, it is scary because you finally feel good in your gut. This is something that my clients report prior to phase 2 and I want to reassure you that even if you react to a specific FODMAP, remind yourself that you can feel good again. It is important that you bring as many foods back in as possible to restore your gut health, so please do not skip this phase. A mantra I'd like to use to ease any anxiety is this: 'What if it all works out?'
- You will be using a tester-food, meaning a food that contains the highest amount of the particular FODMAP we are investigating.
- Challenge with only one FODMAP at a time.
- If you don't get symptoms, work your way up to the third challenge day.
- If you do get symptoms, note them down and do not proceed with the next.

- Once you've completed one FODMAP group challenge, make sure to have a two to three day washout period – that is, strictly low-FODMAP – before starting the next group.

The FODMAP groups

- Lactose
  - o Challenge food: 1.5% cow's milk. Start at ¼ cup and work your way up to 1 cup (250 ml).
- Fructans in wheat
  - o Challenge food: Wholemeal bread. Start at ½ slice of toast and work your way towards 2 slices of toast.
- Fructans in fruit and vegetables
  - o Challenge food: Onion. Start at ½ tablespoon of chopped, cooked onion and work your way towards ½ an onion, cooked.
- Galactans
  - o Challenge food: Chickpeas, lentils or kidney beans. Start at ¼ cup and work your way towards ¾ of a cup.
- Fructose
  - o Challenge food: Honey. Start at 1 teaspoon then work your way towards 2 teaspoons.
- Mannitol
  - o Challenge food: Sweet potato. Start at 75 g and work your way up to 180 g.
- Sorbitol
  - o Challenge food: Avocado. Start with 30 g of avocado and work your way up to 100 g.

Now Table 8 shows an example.

Table 8: A sample FODMAP reintroduction schedule

FODMAP	Day 1	Day 2	Day 3	Day 4	Day 5	Day 6
Lactose (Milk)	¼ cup	½ cup	1 cup	Washout	Washout	Washout
Fructan (Bread)	½ slice	1 slice	2 slices	Washout	Washout	Washout
Fructan (Onion)	½ tbsp.	1 tbsp	½ onion	Washout	Washout	Washout

- If you complete three days with no symptoms, then you are not sensitive to that FODMAP group
- If you react on day 3, then your threshold of tolerance is the portion you've consumed on day 2
- If you experience severe symptoms on day 1 or the next day prior to challenging the second time, then you are sensitive to that particular FODMAP group.

You can then complete the challenges in the same way as the protocol above for the rest of the FODMAPs. It is likely to take anywhere between six and eight weeks to complete the challenges and usually a dietitian would then be able to interpret the results to help you move on to phase 3.

Phase 3: Liberalisation

This is where we celebrate the end of a trying journey investigating what your triggers and tolerances are. My clients send through the results of their challenges for me to interpret and, in return, they receive a 'traffic light' report describing the FODMAPs that they have tolerated (green) and can begin to reintroduce into their diet, those that were tolerated in specific amounts (yellow), and those that they were highly sensitive to (red). (Please note that FODMAP tolerance may change over time. I strongly encourage clients to rechallenge foods they didn't tolerate so well at first testing in a few months' time to explore whether anything has changed.)

Putting it all together

When it comes to managing IBS, there is no doubt that your journey will include addressing the four essential pillars to gaining symptom control and learning to live with this invisible illness.

Now, we've spoken about movement and have gone through different nutritional strategies, but what about the mind and sleep pillars in relation to IBS?

The mind – stress and anxiety

When I was diagnosed with IBS, I thought I had a pretty good grip on my symptoms until I didn't. My missing piece of the puzzle was finding the right approach to manage the mind pillar and I can safely say, that since 2018, which is also known as 'my big mental crash', I have learned to tap into my toolbox of all the approaches and strategies that I've gained through therapy in order to successfully manage, if not 'cure', my IBS.

The two emotional struggles that accompany living with IBS are stress and anxiety, both of which have thrown me into the choppiest waters of my life. I remember hosting a webinar with a therapist on Stress vs. Anxiety and here are some interesting highlights of that talk:

- Stress and anxiety are both emotional reactions that are part of the 'fight or flight' response. With both, we tend to experience very similar mental and physical symptoms, such as a fast heartbeat, rapid breathing, irritability, anger, fatigue, muscle pain and digestive issues.

- Both respond to similar management techniques, yet they are not the same. Stress is a normal, physiological adaptation to change or an external trigger while anxiety happens 'inside us.' Stress is a common trigger for anxiety but anxiety can exist in the absence of stress.

- Anxiety is a sustained mental health disorder that can be triggered by stress. It can affect different areas of our body. It can be characterised by fear and worry, with stress that's out of proportion to the impact of the event.

- The biology of stress affects three body systems in particular: our hormonal (endocrine) system, our immune system and our digestive system.

Given that IBS is a disorder of the gut-brain axis, and knowing how these turbulent emotions cause digestive distress, it is crucial that we act and that we seek support to tackle this axis. I believe that if the mind pillar is neglected, the suffering persists.

What sorts of psychological support exist?

Depending on what suits your situation best, you may need to explore more than one option until you've found the right approach that you are more likely to be consistent with. The list includes, but is not limited to:

- Cognitive behavioural therapy (specifically gut-focused CBT)
- Gut-directed hypnotherapy (GDH)
- Meditation
- Progressive muscle relaxation
- Autogenic training.

In terms of the science, both GDH and CBT have been thoroughly researched providing benefit and relief to IBS sufferers. From personal experience, I've always preferred a combination of approaches, but the true benefit lies in consistency. You're probably thinking, 'What on earth are GDH and CBT'? Before I resume my rambling, here's a quick clarification:

Cognitive behavioural therapy (CBT): CBT is a form of psychological treatment that aims to change unhelpful and ineffective ways of thinking and reacting, allowing the patient to develop better behavioural patterns and coping skills. With CBT, you learn to use more efficient problem-solving skills as well as relaxation techniques to calm the monkey-mind and ease physical tension.

Gut directed hypnotherapy (GDH): GDH is another form of psychological treatment where those with IBS receive repetitive recommendations regarding the control of gut function, while in a hypnotic state, in order to improve their symptoms and psychological state. This involves deepened relaxation techniques and, generally speaking, with 12 GDH sessions, IBS patients experience fewer symptoms, including less gas and bloating, urgency and pain.

Given the safety and favourable outcomes such therapies provide, the mind pillar should be addressed early in the course of managing your IBS and not as a 'last resort' option.

Sandra's confession

IBS is an invisible illness and many living with this condition have suffered a great deal physically, socially and mentally. Pain is a big topic too so here's my confession, and I know many of you might've had a similar experience but wouldn't express this openly.

The last few years have seen me work on my fear of pain. We've been programmed to push through all negative emotions rather than sit and experience the emotion. In the past, the easiest way for me to escape pain was to medicate, using pain-killers, regularly. Too often. Perhaps this also contributed to my dodgy stomach but, at 25, I wasn't sure how to deal with pain.

This is something that many clients of mine have expressed with shame and they end up surprised when I share my own experiences with pain. I am not perfect but I am lucky to be surrounded with people who have made me realise how destructive this can be and it's safe to say that, since starting therapy and being consistent with it for almost five years now, the way I deal with pain, fear, sadness and hurt is very different! I ride that wave, feel every bit of it and shockingly, the turbulences that I used to feel inside are no longer there. The practices that I've learnt have also helped me cope with the insanity of a pandemic.

If you're struggling, please know that there is support out there. I'm lucky enough to work with a network of fantastic therapists whom I call my superheroes. My gut health has improved significantly since I started therapy.

The sleep dilemma

Does poor sleep worsen my IBS symptoms or do my symptoms impact my sleep? Ah, the chicken and egg question when it comes to sleep...

We have very little research to guide us on this specific topic but that doesn't mean we should dismiss the importance of sleep when it comes to IBS management. Poor sleep quality is very common in IBS, possibly resulting from the painful symptoms experienced, such as stomach pain or extremely uncomfortable bloating, which subsequently can cause frequent night waking. The consequences of prolonged sleep deprivation include increased inflammation and stress hormones, in turn, impacting gut symptoms. Perhaps this is why many report having a bad gut after a night of turbulent sleep. It seems like an impossible cycle to break.

Furthermore, just as we have a regulated internal body clock called our circadian rhythm, so does our gut microbiome. If our sleep is disturbed, its natural rhythm is disturbed, possibly jeopardising our gut health further.

While the exact physiological basis around the connection of sleep and IBS remains unclear, the benefits of a good night's sleep are undeniable. For this reason, you'll need to address your 'sleep hygiene' to optimise your gut, mood, immunity and overall wellbeing. Sleep hygiene is just a fancy word that describes the behaviours you engage in to improve the length and quality of your sleep. In simple terms, it just means you are serious about making bedtime a priority.

What changes are crucial to get better sleep hygiene? I've summarised some actions that you can take below, but if you make it to Chapter 12, I've dedicated a more detailed section on achieving a good night's slumber.

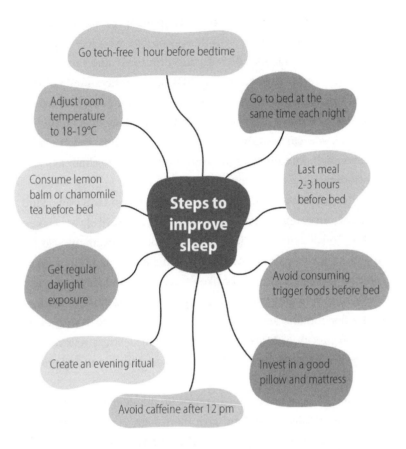

Figure 14: The keys to sleep hygiene

The final word (or more like words really...)

Gaining control of your symptoms and learning to live with IBS is not a straightforward path and will require you to trial multiple approaches until you find 'the one' or 'the many' that work best for you.

Regarding nutrition, it will come down to identifying your individual triggers and that will involve either going through the first-line changes pointed out in this chapter, or specialised dietary approaches. Perhaps your symptoms improve with just eliminating caffeine and alcohol or reducing processed and fatty foods? Otherwise, going through the full FODMAP process may be warranted unless you've identified a pattern after completing a food and symptom diary and noticed that fructans may be a culprit. In that case, you can simply eliminate that group first and check in with yourself. And most importantly, don't forget to challenge and reintroduce rather than eliminate indefinitely.

When it comes to movement, anything is better than none. Find forms of movement that feel good and that you are likely to keep up and be consistent with. Make sure the type of movement also fits in with your reality, so if you decide to join a gym that is located 45 minutes away from home, the likelihood of you showing up is very, VERY slim. Why not look for a group class that is close to work or find a dependable activity partner who you can move with every Saturday morning? Movement should also be a stress-reliever so if it plays the 'stress-inducer' role, rethink your attitude and approach towards it.

As for sleep, identify three habits from the sleep hygiene section that speak loudly to you and give them a go. If you're struggling with an anxious and busy mind in the evenings and find it difficult to fall asleep, then this ritual is for you. I found it extremely useful to come up with a night-time routine that consisted of a cup of chamomile or lemon balm tea and a diary to brain-dump. Keep a journal or a piece of paper by your bedside table. The way I would describe brain-dumping is writing down every thought and feeling that comes to mind, think of it as 'word vomit' on paper. Don't worry if what you've noted down doesn't make sense! Just keep on going until you feel you've 'cleared' your mind. Dumping all these thoughts on a piece of paper right in the evening has helped me tremendously with my own anxiety and sleep.

And finally, if there's one thing I have encouraged all my clients to do at the start of IBS therapy, it is to tackle the mind. We can never be rid of stress but it comes down to learning how to cope and being consistent with our coping strategies to become stress resilient.

Flare ups are a fact of living with IBS

Before I end this chapter, I wanted to add my five coping strategies if you do experience a flare up of your symptoms. Accepting that flare ups happen occasionally makes encountering them a little more tolerable. You might be going through a difficult time and then, bam! You're rushing to the porcelain throne, questioning life. The bloating and cramping have brought you back to where it all started and you can't help but wonder – 'WHY?'.

Coping tip #1: Keep calm

Reflect by journaling or noting down in your food and symptoms diary what you've consumed, including your stress levels, to check which potential triggers might have caused this.

Coping tip #2: Go back to LOFO eating

You may need to revisit the low-FODMAP phase only for a couple of days until your symptoms settle. Give your gut time to recover before bringing your usual foods back.

Coping tip #3: Include a ton of self-care

This may come in the form of resting on the couch with a hot water bottle on your tummy. You may also include some form of meditation or belly breathing as well as some gentle yoga.

Coping tip #4: Be prepared

Make sure you have the following stocked up in an allocated 'flare up' drawer or cabinet: ginger or peppermint tea, peppermint oil capsules, low-FODMAP snacks as well as probiotic and fibre supplements that are symptom specific.

Coping tip #5: Seek help when in doubt

You may need to revisit your GP, gastroenterologist, dietitian or even therapist, and that's absolutely fine, especially if you've been overwhelmed.

10

The one that's all about food intolerance

'I am allergic to gluten and dairy.'

Those are the famous few words that have transformed millennials into gluten-loathing, celery-loving, wellness enthusiasts. Don't get me wrong, many people can actually be allergic to wheat and cow's milk protein, but there's a huge lack of understanding and a likelihood of misdiagnoses, especially when social media and wellness unicorns are the go-to source for health-related answers.

Food allergies and intolerances are becoming increasingly common and the rise of the self-diagnostic, pseudo, unorthodox tests tends to follow. With all the misinformation out there, people tend to self-diagnose or claim to have a specific 'allergy' or 'intolerance', with a huge lack of understanding of what either really is. So, is it a food intolerance or a food allergy that you may be suffering from?

Many of us suffer unpleasant reactions and symptoms after consuming certain foods, causing us to think that we are 'allergic' to that food. In most cases, you might be intolerant; however, it all depends on whether your immune system is involved or not. Understanding the difference between food intolerance and a food allergy is essential as diagnosis and dietary management differ for both.

Allergy vs intolerance

Despite dedicating this chapter to food intolerances, we have to mention allergies. So, what is a food allergy? In simple terms, a food allergy is your immune system's reaction against a specific food protein, where it responds to that food as a harmful substance. This response involves the release of allergy antibodies called IgE antibodies against that particular protein. Allergic reactions would then follow and can manifest in many ways, from minor symptoms such as a rash, itching or hives to more serious reactions such as swelling of the tongue, tightness of the throat and breathing difficulties. For this reason, a food allergy can be fatal. The most common foods people are allergic to include:

Eggs

Cow's milk

Peanuts

Tree nuts (cashews, walnuts, hazelnuts and almonds)

Fish and shellfish

Soya

Wheat.

Food allergies are common in children and they tend to grow out of these; however, some allergies – such as seafood and nuts – can be life long. Adults can also develop allergies against certain fruit and vegetables and this is known as an oral allergy syndrome. Common food allergens in adults include tree nuts and stone fruit such as plums and apricots, to name a few.

Diagnosing a food allergy involves skin prick tests or blood tests for allergen specific IgE (RAST), which will help your doctor confirm which allergens you are sensitive to. It is important to note that allergy test results cannot be used on their own and must be considered together with your medical history. I could devote a whole book to talking about food allergies, given the complexity of it and the variety of allergies that exist, but I want to talk here specifically about the rise of food intolerance given that it appears to be increasingly prevalent, so here goes.

With an intolerance, the immune system is not involved (see my warning below about bogus 'IgG testing' for food intolerances in the Box) and no IgE antibodies are produced against the food consumed. By definition, a food intolerance is 'a non-immunological response initiated by a food or food component at a dose normally tolerated and account for most adverse food responses'. Basically, a food or food component you've consumed caused a chemical reaction that did not involve your immune system but irritated your gut and perhaps your nervous system too.

Food intolerance is estimated to affect about 20% of us humans yet it is complicated, poorly researched and a maze to navigate. To complicate things further, there are various types and symptoms that are not only gut related. This makes my job FUN. Actually, working in the field of food intolerance is not only challenging but extremely fascinating as it takes the saying 'There's no one-size-fits-all-approach' to a whole new level.

What then is a food intolerance?

The most common food intolerances can be divided into three main types:

1. Enzymatic, such as lactose intolerance where the enzyme lactase is deficient

2. Food chemical (also referred to as a pharmacological sensitivity), such as to natural food chemicals (e.g. amines) or additives (e.g. sulphites)

3. FODMAP sensitivity (we've covered that in great detail already).

As for symptoms, take a look at Figure 15.

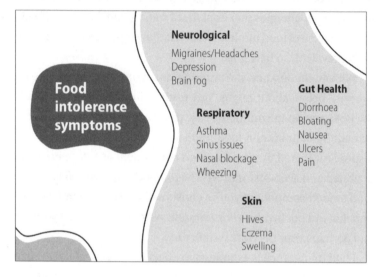

Figure 15: The symptoms of food intolerance

It is important to note that medical advice is recommended to eliminate other causes of the symptoms shown in Figure 15. Food intolerance can be a possibility worth investigating if there is no established medical cause for such symptoms.

Why do we develop a food intolerance?

The exact mechanisms are still under review, especially when it comes to food chemical intolerances (see page 174), but the possible causes include a metabolic or a digestive disorder (e.g. IBS) or picking up an infection, such as food poisoning, that has negatively impacted the integrity of your gut lining, causing a rise in food sensitivities. Also, age is a factor – as we get older, our digestion slows down and enzyme levels, such as lactase, decrease, increasing our risk of developing an intolerance. As we age, we may become more sensitive to certain food chemicals and additives, such as sulphites in wine and dried fruit, or monosodium glutamate (MSG), commonly found in ultra-processed foods and spice mixes.

The biggest food intolerance myth

You should have a 'food intolerance test' to identify what you're intolerant to.

Fact: There are no blood tests that reliably identify a food intolerance.

A large number of so-called intolerance tests have inundated the healthcare scene with false claims of diagnosing food intolerances. First of all, contrary to the practice of some medical, alternative and natural therapy clinics, there are no blood tests that reliably identify food intolerances. These tests are usually very expensive and often indicate a very long list of trigger foods to be avoided. The most common unorthodox tests of allergy and intolerance are those that are based on IgG food antibody testing. IgG antibodies are proteins produced by the immune system in response to exposure to external triggers, like pollens, foods or insect venom. IgG antibodies to food are commonly detectable in healthy adult patients and children, whether food-related symptoms are present or not. So, IgG antibody testing simply indicates food exposure and not necessarily 'intolerance'. In fact, higher levels of IgG to foods may simply mean more tolerance to those foods. There is no credible evidence that measuring IgG antibodies is useful for diagnosing a food allergy or intolerance, nor that IgG antibodies cause symptoms. Despite studies showing the uselessness of this technique, it continues to be promoted.

Unfortunately, the only way to identify a food intolerance is to undergo an elimination diet under the supervision of an accredited dietitian or registered nutritionist. However, if you do choose to get tested for an allergy or intolerance, don't be afraid to ask the practitioner these questions:

- Is there scientific evidence that it works? Has such evidence been published?

- How much does it cost?

- Why doesn't my own doctor suggest this type of treatment?

- What are the qualifications of the practitioner recommending the treatment?

What are the 'immunology officials' saying about all these sketchy tests? Well, if there's one thing that the Australasian Society of Clinical Immunology and Allergy, the Canadian Society of Allergy and Clinical Immunology (CSACI) and the British Society for Allergy and Clinical Immunology are in agreement with it is that there is no credible evidence to date to support the use of measuring IgG antibodies to diagnose an allergy or intolerance. They strongly discourage their use and have described them as being invalid, irrelevant and unreliable.

Why are they still being promoted you may ask? Well, there's money to be made from vulnerable and desperate people. It is unfortunate, but the more that these tests are promoted and used, the more money people make and that also includes doctors and even dietitians who generate a certain revenue with every sale. Ethical? Absolutely not. Does it happen? Yes.

How is food intolerance diagnosed?

Regrettably, not as quickly and straightforwardly as you would've hoped since you'll need to undergo an elimination diet. Yes, we've come across something similar when we addressed the low-FOD-MAP diet. The same principal applies to food chemical intolerances and other types as you'll read. The elimination diet will have varying degrees of restriction depending on your case and is followed for a minimum of three weeks. Subsequently, you'll enter the challenge phase where each suspected food or chemical is introduced and challenged one at a time. If symptoms reoccur after a food is challenged, then it is likely that the substance introduced was responsible for triggering the symptoms.

When it comes to the different types of food intolerance, I want to highlight the most popular ones but not to fail to mention that a sensitivity to FODMAPs is also considered a non-immunological food intolerance. The reason I'm mentioning FODMAPs is due to the fact that many who claim that they are intolerant or 'allergic' to gluten, are intolerant to the fructans (FODMAPs) found in wheat. Let's take a deep dive and look at the most popular type of intolerance, and that is to wheat.

Wheat – Non-coeliac wheat sensitivity (aka non-coeliac gluten sensitivity or NCGS)

I want to start off by explaining the different components of wheat that are problematic:

- Gluten
- Fructans
- Amylase-trypsin inhibitors (ATIs).

Keeping the level of complex science jargon to a minimum, gluten is the main protein found in wheat, fructans are the fermentable carbohydrates (FODMAPs) and ATIs are also a type of protein. Gluten and ATIs play a huge role in the development of coeliac disease (an autoimmune condition where a lifelong elimination of gluten is crucial – see page 118) as well as wheat allergies. Then, there's a whole world of gluten-related disorders for which there are no conclusive diagnostic tests and unexplained mechanisms by which wheat causes problems. Enter non-coeliac wheat sensitivity (NCWS) or, as many refer to it, 'wheat intolerance'.

Firstly, those of you who are quick to blame gluten as the cause of your gut woes, think again as we now know that there are other elements in wheat that may be to blame. NCWS is a recent entity that was possibly born out of a small percentage of people who did not fall under the category of being sufferers of coeliac disease or a wheat allergy. The actual mechanisms involved in developing NCWS are still under review, where ATIs play a more dominant role in eliciting symptoms than gluten. Hence, the recent name change from non-coeliac gluten sensitivity to non-coeliac wheat sensitivity made more sense.

What are the symptoms of NCWS?

To keep things as confusing as humanly possible, symptoms are similar to those of IBS and include extra-intestinal ones. They generally occur shortly after ingesting wheat and can last from a few hours to a few days:

- Abdominal pain
- Bloating
- Diarrhoea, constipation or both
- Nausea
- Headache

- Fatigue
- 'Foggy' mind
- Rash
- Disturbed sleep
- Joint pain.

When other symptom manifestations include your skin and/or nervous system, we can generally assume that FODMAPs are not a culprit and that we're dealing with other problematic components. In terms of coming to a diagnosis of NCWS, we must exclude coeliac disease and wheat allergy first. There are no validated diagnostic tools, blood tests or intestinal biopsies that can confirm whether we are dealing with NCWS except for undergoing an elimination diet of wheat, seeing a positive response on a wheat-free diet and reacting to challenges when wheat is reintroduced.

The 'nocebo' effect

Gluten has been demonised, gaining an undeserved bad reputation throughout the last decade. A gluten-free diet has risen as the saviour of all things related to gut-ailments and beyond. The issue that we are seeing with wheat sensitivities is that most people with suspected NCWS may be to some degree influenced by a 'nocebo' effect when they commence challenging wheat i.e. gluten. The negative placebo effect, nocebo, is when a person is already expecting to experience unpleasant symptoms going into a challenge. This is because they already have a biased expectation about the consequences of consuming gluten.

'I am intolerant to gluten'

These are the famous few words that come out of every other person complaining of gut woes. Who should actually avoid gluten then?

- Those diagnosed with coeliac disease: This is an auto-immune disorder where gluten causes damage to the lining of the small intestine.

- Some diagnosed with Hashimoto's thyroiditis may benefit from a gluten free diet.

- Those with non-coeliac wheat sensitivity (NCWS) but research is still inconclusive as to whether gluten is the actual culprit causing symptoms in the absence of coeliac disease.

- People with IBS following a low-FODMAP diet eliminating fructans in wheat but noting that not all gluten-free products are low FODMAP.

Food chemical intolerance

Food chemical intolerance, also known as a 'pharmacological food intolerance', refers to a sensitivity towards food chemicals that are either naturally occurring, such as amines and salicylates, or added during food processing, such as glutamates, benzoates and sulphites. The mechanism by which one develops a food chemical intolerance is confusing and poorly understood, but it is believed that such chemicals may act as irritants on the nervous system.

The most common symptoms of a food chemical intolerance include chronic hives, low blood pressure, headaches or migraines, asthma and rhinitis, as well as digestive symptoms such as stomach pain, bloating and diarrhoea. Once again, we are left with a wide spectrum of manifestations that are not only frustrating for the one experiencing them, but also for a dietitian like me, playing detective until we get to the bottom of the problem.

And just to make things even more puzzling, the level of sensitivity to a particular food chemical may also vary at different times depending on factors such as stress, hormonal cycles, exposure to environmental chemicals and use of medications.

Table 9 lists some examples of foods containing natural and added food chemicals thought to trigger both gastrointestinal and extra-intestinal symptoms.

Table 9: Foods that may trigger chemical sensitivities

Food chemical	Food source
Amines (e.g. histamine)	Chocolate, cheese, cured meats, smoked fish, avocado, tomatoes, wine, beer, spinach, bananas…
Salicylates	Apples, tomatoes, herbs and spices, peppermint, nuts, wine, beer, honey…
Sulphites	Wine, dried fruit, sausages, soft drinks, cordial…
Monosodium glutamate	Commercial foods such as savoury sauces, all spice mixes, snack foods, Chinese fast food, some soy sauces…
Nitrates	Preservative commonly found in cured/processed meats such as ham, bacon, salami and other cold cuts…
Benzoates	Preservative found in soft drinks, cordials, commercially packed dips, fruit juices…
Propionates	Preserved breads and other packaged, baked items
Artificial colours	Confectionary, commercially baked and packed cakes and muffins, soft drinks, jelly…
Synthetic antioxidants	Fried snack foods, margarine…

Histamine intolerance

I can guarantee that you've encountered 'histamine intolerance' in one of your searches of the internet about 'healing your gut' and it is becoming more prominent than we realise. Histamine intolerance, also known as 'enteral histaminosis' or sensitivity to dietary histamine, is a disorder associated with an impaired ability to break down ingested histamine. Histamine is a chemical (or, in scientific terms, a bioactive amine endogenously produced by its precursor amino acid, histidine) that plays a huge role in our immune system. Histamine is produced by our bodies but can also be found in food.

Why would someone develop an intolerance to histamine? There are numerous reasons why and the most common include medications that block the production of the enzyme diamine oxidase (DAO), such as antibiotics, inflammatory bowel disease, high histamine foods and foods that trigger the release of histamine and SIBO.

Although known since the beginning of the 21st century, interest in histamine intolerance has recently grown due to the lack of research when it comes to diagnosis and its clinical management. (Yes, I am very aware of how repetitive I am starting to sound when it comes to 'lack of research' but that's science – or lack of it – for you!) The main cause of exogenous histamine intolerance is a deficiency in DAO or a possible defect in its functioning. Histamine is present in a wide range of foods in highly variable concentrations. Examples that contain high levels include preserved or processed meats, fermented products, citrus fruit, cocoa and vegetables such as spinach, aubergine, tomatoes and avocado.

Symptoms of histamine intolerance generally occur during or immediately following meals and include the following:

- Dermatological: itchy skin, sudden reddening of the skin (flush symptoms), angioedema (swelling under the skin)
- Neurological: Migraines or headaches

- Digestive: Vomiting, nausea, diarrhoea, stomach pain
- Respiratory: Constant sneezing, nasal drip
- Circulatory: Low blood pressure, dizziness and having a fast heart rate.

Diagnosing histamine intolerance involves the presence of two or more of the typical symptoms mentioned above as well as seeing an improvement with a low-histamine diet and/or anti-histaminergic medication.

Oral supplementation with DAO enzyme is also being used to help the intestine to degrade dietary histamine. A common supplement used in this case is DAOSIN.

The low-histamine diet is another elimination diet that is specific to histamine. It requires you to eliminate high-histamine foods for two to four weeks followed by a reintroduction/challenge phase to establish your histamine threshold. By now, you'll be pretty much getting the hang of elimination-diet-talk where, I repeat, you will need to go through the reintroduction phase as fearful as it may be. This is why I highly recommend finding a specialist who can be your ally throughout this journey.

The low-histamine diet is progressively gaining evidence to support its clinical efficacy but there is a lack of unanimity on foods that must be avoided. Furthermore, there is a ton of conflicting information out there. The Swiss Interest Group Histamine Intolerance (SIGHI) have the most up-to-date and comprehensive food list, which I have personally used with my clients. However, below is a summary of the foods most experts agree on that need to be limited:

- Fermented foods: fermented dairy products (yoghurt, aged cheese and quark), fermented soya products (miso, tempeh, soy sauce), fermented or pickled vegetables.
- Tinned, canned, cured, smoked, processed fish and meat.

- Fruit: citrus fruit, strawberries, pears, bananas, grapes, avocado, raspberries and all their fruit juice.

- Vegetables: spinach, aubergine, olives, tomatoes and tomato products.

- Stock (bouillon), broth and vinegar.

- Chocolate, cocoa.

- Green tea, matcha and coffee.

- All alcohol.

- Ultra-processed foods and junk foods that contain artificial colours and flavourings.

Are there any supplements that can help with histamine intolerance?

The big three include vitamin C, vitamin B6 and some probiotic strains. The research is still scarce, but vitamins C and B6 are important for DAO to function properly, in turn acting as antihistamines themselves given their support for histamine breakdown.

When it comes to probiotics, the strains that have shown interest are mainly those that were seen to help in preventing allergic reactions and have the potential to minimise histamine release. Given that fermented foods (natural and rich sources of probiotics) are off the table, it may be worthwhile considering a probiotic supplement; the researched species include *Lactobacillus rhamnosus, Bifidobacterium infantis* and *Bifidobacterium longum*. You need to be careful with probiotic supplementation as some strains are also known to aggravate histamine production and they include strains of *Lactobacillus casei* and *Lactobacillus bulgaricus*.

The low food-chemical shopping list (Strict)

This shopping list serves as a starting point if you wish to trial a strict low-chemical diet for two to four weeks. The elimination phase is known as the RPAH Elimination Diet or The Failsafe Diet, developed by the Allergy Unit at Royal Prince Alfred Hospital in Sydney, Australia. There are differing levels of restriction depending on the severity of symptoms and access to the full elimination protocol is widely available. The following list serves as a snapshot:

Grains and cereals
(Wheat-free)
  Amaranth
  Buckwheat
  Gluten-free bread made
  from rice or buckwheat
  Gluten-free Weetabix
  Millet
  Noodles (based on rice,
  buckwheat, legume, konjac)
  Oats
  Quinoa
  Rice (white and brown)
  Sago
  Sorghum
(Wheat/gluten-containing)
  Barley
  Couscous
  Plain biscuits
  Rye

Semolina
Wheat crispbreads
(unflavoured)
Wheat flour
Wheat pasta/noodles
Fruit and vegetables
  Bamboo shoots
  Beans (green, yellow,
  French or string)
  Brussels sprouts
  Cabbage (red, green or savoy)
  Celery
  Garlic
  Iceberg lettuce
  Leek
  Mung bean sprouts
  Pears (fresh, ripe, peeled,
  canned in syrup)
  Potatoes (peeled)
  Swede

Meat and meat alternatives (protein)
Beans (borlotti, cannellini, lima, butter, red kidney)
Beef (fresh and not aged)
Calamari
Chickpeas
Crab
Fresh eggs
Lamb
Lentils
Lobster
Mussels
Oysters
Sea scallops
Skinless chicken
Split peas (green and yellow)
White fish

Nuts and seeds
Cashews (raw or lightly roasted)
Poppy seeds
Pure cashew butter

Dairy
Butter
Cheese (ricotta, cottage, cream cheese, mascarpone, quark)
Cream

Dairy milk (fresh, powdered, UHT, canned, plain)
Sour cream
Yoghurt (plain, natural)

Soya-based foods
Plain tofu
Plain/unflavoured soya milk
Soya cream cheese
Soya yoghurt (natural)
Soya ice cream (vanilla)

Cooking and baking
Canola-based spray oil
Carob powder
Chives
Citric acid
Garlic
Margarine made with sunflower, safflower, canola or soya without antioxidants, preservatives or annatto
Oils (not cold-pressed): Rice-bran, canola, sunflower, safflower, soya oils without antioxidant
Parsley
Saffron
Salt
Vanilla essence or pods

Beverages
Decaf coffee
Plain whiskey, gin, vodka
Rice milk (plain, vanilla –
ideally with added calcium)
Water.

Where do I start?

This is probably the biggest question you are faced with if you are trying to figure out which food chemical to avoid or, indeed, if you are suffering from a food intolerance.

First and foremost, avoid self-diagnosing or consulting Google or any other form of social media. I would then suggest you start tracking the following for a minimum of seven days to a maximum of two weeks:

- Foods consumed in detail
- Beverages consumed in detail
- Symptoms
- Timing of symptoms
- Activity
- Stress (use a simple scale from 0-10 where 0 = no stress, and 10 = extremely overwhelmed).
- Sleep (note down hours of sleep and whether you felt rested in the morning)
- Stool
- Menstrual cycle (if relevant).

Once you've completed your tracking, ask yourself the following:

- Are you seeing patterns?
- Is stress, sleep or mood influencing your symptoms?
- Is a single food causing your symptoms?
- Multiple foods?

What I generally advise when it comes to suspected food intolerance is to seek the following specialists:

- Allergist/immunologist
- Gastroenterologist
- Clinical dietitian or registered nutritionist with experience in gastroenterology and food intolerances.

You may need to rule out other medical reasons for your symptoms first before suspecting a food intolerance. The elimination phase when managing food intolerance is essential to help you achieve symptom control, followed by identifying your triggers and threshold when you move into the challenge phase. This whole process is always recommended to be done with the support of an experienced dietitian or nutritionist to prevent unintentional weight loss, malnutrition, disordered eating, fear of food and poor diversity, all of which are far too common in people struggling with a suspected intolerance and unexplained symptoms. The easiest way to find a specialist is to ask your GP, who will be able to refer you to one of the health professionals mentioned above who will help you manage a food intolerance and navigate any change you will need to embark on to gain symptom relief.

11

The one on bum-dumplings – the world of haemorrhoids

'THIS IS NOT HAPPENING,' Sienna silently squealed as she sat on the toilet annoyed while feeling a couple of tiny, bulging bum-dumplings making an unwelcomed arrival on her holiday. She realised that all the straining had taken a toll on her derrière and that 'traveller's constipation' was proving to be unfixable. As she emerged from the bathroom, she melodically called out to her partner, Lee, 'Darling, I have a huge favour to ask. I need you to find the closest pharmacy and ask them if they've got any quick fixes or creams that can help with...' she paused for a second and whispered, 'haemorrhoids.' She paused. 'Yes, I said it and I am fully aware of how gross this is but I just want to sit on a chair with ease without feeling like my bum is creating extra cushioning and I'm about to ascend towards the ceiling!' And off Lee went in search of a pharmacy to ease his partner's bum pain.

As I begin to write this chapter while in Italy, I can't help but think of how many holiday-makers currently feasting around me are constipated and share Sienna's bum struggles. Actually, I can't help but wonder what the hotel staff around me think as they take a glimpse at my open chapter and articles entitled 'bum-dumplings', 'piles' and 'haemorrhoids'.

Haemorrhoids, or piles, are more common than we think, but the exact prevalence is unknown. Perhaps this is because we're not vocal about them until they become worrisome and unresponsive to Google's home-remedy suggestions.

What are haemorrhoids?

Haemorrhoids are swollen or inflamed veins around your anus or lower rectum. In simple terms, they are either internal or external depending on where they are. External haemorrhoids (what most us associate the term with) are the most bothersome and form under the skin around your anus. Internal haemorrhoids may go completely unnoticed due to their location or can prolapse (protrude out of the anus) depending on how inflamed and swollen they become, stretching towards the 'light at the end of tunnel'.

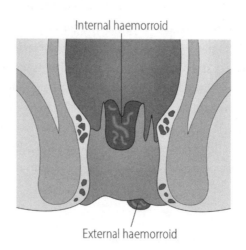

Figure 16: Internal and external haemorrhoids

When people ask about the signs and symptoms of haemorrhoids, my answer in brief is 'you'll know'. The most common symptoms include:

- Painful lumps felt around your anus
- Anal itching
- Anal pain, especially when sitting down
- Blood in your stool or on toilet paper after a bowel movement
- Painful bowel movements.

Now, it is important to address these symptoms with your doctor because not every anal symptom is due to haemorrhoids. These symptoms may also be a sign of other conditions such as inflammatory bowel disease or forms of colon or rectal cancer. Rest assured though that haemorrhoids are actually the most common cause of anal symptoms but, as always, make sure you seek support to rule out other causes. If blood is involved, don't wait or take guesses. See your doctor or specialist.

Causes and diagnosis of haemorrhoids

What causes haemorrhoids?

- Chronic constipation or diarrhoea
- Pregnancy and child birth
- Straining
- A low-fibre diet
- Heavy lifting
- Sitting on the toilet for a long time
- Weakening of the tissues around your anus and rectum that occurs with ageing.

Generally, your GP or specialist can diagnose haemorrhoids by discussing your medical history and performing a physical examination, also known as a digital rectal exam. In some cases, they may ask to perform a sigmoidoscopy to look for internal haemorrhoids.

Once a diagnosis has been made, your haemorrhoids are then graded from I – IV:

Grade I: No signs of prolapse

Grade II: Prolapsed but retract on their own

Grade III: Prolapsed but can be manually pushed back in

Grade IV: Prolapsed but cannot be pushed back in without a lot of pain.

Anal glossary

Since we're exploring the land deep down under, you may come across some unfamiliar terms:

Anal fissure: Sometimes confused with haemorrhoids due to similar symptoms, anal fissures are tears along the tissue lining of the anus. They can occur if you pass dry, hard stool (i.e. if you are constipated) but also during childbirth or anal intercourse.

Skin tags: A skin tag that forms around the anus is just the extra skin left behind after a blood clot in an external haemorrhoid has resolved.

Thrombosed haemorrhoid: This is when your haemorrhoid develops a blood clot in a haemorrhoidal vein, causing an obstruction in blood flow. Thrombosed haemorrhoids can be very painful and swollen and may cause rectal bleeding.

The prevention plan

Before we talk about treatment options, I want us to dive into PREVENTION first. Obviously, ageing is a fact of life that we cannot control and haemorrhoids are common as we get older. However, what we can influence are crucial areas that lower our risk of developing bum-dumplings. The key areas to address include managing constipation, aiming for regular bowel movements without any straining and consuming a diet that has the right amount of fibre and plants to support your bowel movements. The six-step plan that's part of the constipation toolbox on page 103 applies to haemorrhoid prevention. There's no point reinventing the wheel hence Figure 17 serves as a gentle reminder to keep this toolbox in mind.

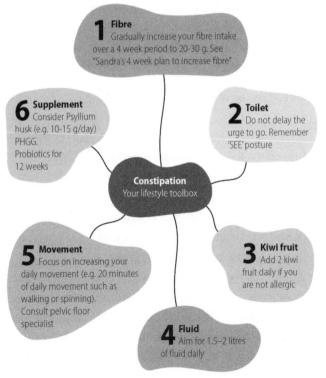

1 Fibre
Gradually increase your fibre intake over a 4 week period to 20-30 g. See "Sandra's 4 week plan to increase fibre".

6 Supplement
Consider Psyllium husk (e.g. 10-15 g/day) PHGG. Probiotics for 12 weeks

2 Toilet
Do not delay the urge to go. Remember 'SEE' posture

Constipation
Your lifestyle toolbox

5 Movement
Focus on increasing your daily movement (e.g. 20 minutes of daily movement such as walking or spinning). Consult pelvic floor specialist

3 Kiwi fruit
Add 2 kiwi fruit daily if you are not allergic

4 Fluid
Aim for 1.5–2 litres of fluid daily

Figure 17: The six-part plan for haemorrhoid prevention

Sienna's holiday struggles are far too common so here's a quick rundown of how to prevent traveller's constipation:

- Psyllium husk can be your travel companion depending on whether food items are allowed at your destination. Take 1 teaspoon daily dissolved in a glass of cold water followed by another glass.
- Ensure you are well hydrated. Aim to consume 2 litres of water daily.
- Find the kiwi fruit at the breakfast buffet.
- Prune or pear juice can do the trick if you're really clogged up.
- Include at least two different vegetables at lunch and dinner.

When it comes to lifestyle behaviours in preventing hae-morrhoids, a lot will involve your toilet habits, hygiene and properly-fitting underwear. The four toilet habits I urge clients to be mindful of are:

1. Never delay the urge to go.

2. Do not strain.

3. When on the toilet, remember to SEE (page 105).

4. Avoid sitting for too long.

Delaying the urge to go and holding your poo in on occasion won't cause you much harm. However, making it a habit to hold in your stool can cause unwanted health effects. For example, holding in a stool for too long can cause it to turn hard, which makes going to the bathroom difficult and uncomfortable. This in turn could lead to constipation, haemorrhoids and, in some instances, anal fissures. The other consequence of constantly delaying the urge to poo is something called rectal hyposensitivity. Nerve damage around the rectal area occurs as does a glitch in how your colon and gut communicate, causing you to no longer feel the urge to go.

Since the cause of haemorrhoids involves placing a lot of pressure on the blood vessels in the anus and rectal region, straining is a habit that should be avoided. The SEE posture creates an ideal position for your pelvic floor muscles to evacuate stool without the need for straining so make SEE a habit.

Finally, there's no need to reiterate the importance of bum hygiene but here goes: please wipe well! If you do have a case of bum-dumplings, you can opt for wet wipes to gently pat and clean your bum after a poo. You can also use a tissue moistened with water, then gently pat your bum dry.

The treatment plan

Your treatment plan will depend on the severity of your haemorrhoids. Generally, low grade haemorrhoids will resolve on their own within a week or so. Similar to prevention, first-line therapy involves addressing fibre and fluid consumption, including the use of stool softeners such as psyllium husk. The goal is to keep your stools soft to avoid irritating and worsening your bum-dumplings. You will also be instructed to avoid straining and to keep your phone out of the bathroom to limit your time on the toilet.

When it comes to increasing your fibre intake, remember to do so gradually (e.g. in 5-gram increments). Foods with ≥3 grams of fibre per 100 grams or ≥10% of daily reference value per serving are considered a 'good source of fibre'. If a food has twice as much fibre, it is deemed 'high fibre'. If you're not sure where to start, Table 10 lists food sources including their fibre content.

Table 10: The fibre content of common foods

Food	Fibre content per 100 grams (g)
Wholemeal bread (2 slices)	7 g
Oats	8 g
Pearl barley (cooked)	4 g
Wholegrain pasta (cooked)	4 g
Figs	7 g
Kiwi fruit	3 g
Prunes	7 g
Avocado	6.7 g
Broccoli (cooked)	2.8 g
Green peas (cooked)	5.6 g
Parsnips	3.5 g
Carrots	2.8 g
Almonds	6.8 g
Peanuts	7.6 g
Sunflower seeds	5 g
Lentils	7.3 g
Chickpeas (canned)	6.7 g
Edamame (cooked)	5 g
Hummous (plain)	6 g

Other home remedies can include the use of over-the-counter creams and suppositories that contain lidocaine or other compounds called phlebotonics that offer relief. Your doctor may also prescribe NSAIDs (non-steroidal anti-inflammatory drugs such as ibuprofen) for pain relief and easing the inflammation.

A sitz bath with Epsom salts twice a day may also offer some bum relief and if you're unaware of what that is see Figure 18.

Figure 18: Sitz bath

What are phlebotonics?

Indeed a mouthful, phlebotonics are a class of medication of plant origin that are believed to improve the health and strength of blood vessel walls. They have been prescribed as part of haemorrhoidal treatments as studies have shown a beneficial effect in treating symptoms without any significant adverse effects. Phlebotonics include plant flavonoids, which are commonly known as plant pigments with anti-inflammatory, antioxidant properties.

A few examples of these plant extracts include:

- Diosmin, which is a natural extract found in citrus fruit peel
- Troxerutin, extracted from the Japanese plant *Sophora japonica*
- Rutin, commonly found in buckwheat, black tea, green tea and elderflower
- Hesperidin, found in citrus fruit
- Quercetin, found in onions and apples
- Horse chestnut seed extract, derived from 'conkers.'

Most haemorrhoidal phlebotonic supplements are a mixture of different extracts but, despite being of plant-origin, there are always contraindications such as pregnancy, breastfeeding and medical conditions such as kidney disease.

If you've applied all the above but conservative treatment has failed and symptoms do not improve after a week, it is important you seek support from your specialist to discuss next steps. For more serious cases, there are interventions one can consider and they include things like rubber band ligation, infrared coagulation, electrocoagulation and sclerotherapy. These procedures can be performed in your specialist's clinic; however, there are surgical interventions that may be warranted depending on the grade and degree of severity of your haemorrhoids. A haemorrhoidectomy or haemorrhoid stapling are performed on haemorrhoids that do not respond to all the treatments mentioned above.

I don't want to go into too much detail when it comes to the procedures but I do want to highlight that you will have to apply the same principles post-surgery that are included in the six-step plan depending on where you are in your recovery journey.

Ending the stigma

Having to go on with your day to day life with a painful and itchy bum can definitely take an emotional toll, especially if your haemorrhoids are large and have significantly prolapsed. While we find a lot of humour in talking about all things 'anus' and 'rectal', I know you won't be surprised to hear that it can significantly impact a person's quality of life, leaving them with a feeling of embarrassment and shame. As a matter of fact, many people avoid seeking treatment for fear of embarrassment.

If you have any hesitation about seeking medical support, keeping the following points in mind may help:

- Your doctor has seen it all and has probably come across worse cases. They have seen a ton of haemorrhoids across their career given how common the problem is.

- Admit to your doctor that it is difficult for you to talk about it and openly express your embarrassment and hesitancy to seek support. Your doctor will put you at ease and will be considerate of your feelings.

- Know what to expect. I think it is important to either call or email in advance to ask about what to expect at your appointment to avoid any surprises. Your appointment will most likely include a physical exam so be prepared to strip down for your doctor to either have a visual examination or they may need to insert a small scope into your anus to have a quick look for internal haemorrhoids.

Voicing your pain to the people closest to you is something I can recommend as you'll be surprised how many within your social circle have struggled with haemorrhoids. From personal experience, when I had a case of bum-dumplings after giving birth, little did I know that two of my closest friends had much more serious cases, one requiring hospitalisation and surgery.

While researching haemorrhoids, I came across a magazine article claiming that one in three women will not talk about the following issues, dubbing them 'too awkward to talk to my friends or doctors about':

- Haemorrhoids
- Bloating
- Gas
- Constipation
- Thrush

- Acne
- Excessive sweating.

This concerned me and here's why:

1. If we don't speak up, we might be missing a serious diagnosis that can have serious consequences. For example, chronic, painful bloating is not normal; it can point to a whole bunch of gut disorders. What many of you are inclined to do is hop onto the internet, search symptoms, and fall into the rabbit hole of trial and error alone, trying to get to the bottom [ho! Ho!] of it. The worst part of it all is your falling victim to all the pseudoscience out there and not getting support from qualified, experienced, health professionals.

2. The other thing that has been brought to my attention via my own experience and through that of my clients is that as women, we are very good at 'soldiering on' and the 'pushing through' approach. 'This is my norm' and I should just live with it. Unfortunately, this is detrimental to our mental and overall health since we end up bottling up our struggles and suffering in silence.

My request is for us to break the taboo around topics involving our health, our bums and the land down-under and get checked when something is NQR (not quite right!).

12

The one that ends with a blissful gut

The journey towards a blissful gut is a turbulent one, if not one that puts your mental wellbeing to the test given the frustrations you will encounter. However, do not give in to the idea that it is unattainable.

Depending on what condition you suffer from or have experienced, pursuing a blissful gut will require you to put the pieces of your own gut puzzle together, and each of the four pillars of gut health that we've gone through will have its respective role to play. Actually, what many fail to mention out loud as we strive for equality amongst all sexes is that when it comes to our gut, there are significant gender differences.

The science continues to reveal how males and females demonstrate important variations when it comes to our gut microbiome and that the main driver for these differences are hormones.

The female gut

There is no doubt that women have been underrepresented in clinical trials historically (especially women of colour) so I am dedicating this section to the female gut and what we currently know. (Well, if you consider how we are an ever-changing mystery, good luck to science in trying to uncover the full picture!) This will be a condensed section since I believe the female gut warrants a book in itself.

Why highlight this sex-variation you may ask? Sex hormones are believed to be a driving force when it comes to the prevalence and severity of medical conditions – for example, autoimmune conditions and IBS are more common in females. Furthermore, even symptom type and severity can differ between men and women, such as in the case of IBS. Females tend to have increased constipation in comparison to men, possibly due to the higher levels of progesterone and oestrogen. Progesto what?!

When discussing hormones and our gut, you'll need a quick rundown so here's my version of a speedy 'hormone glossary':

- Hormones: Chemical compounds or 'messengers' produced by the endocrine system (i.e. our body's communication network) that control many bodily functions and processes, including growth, reproduction, metabolism, sleep and appetite, to name a few.

- Oestrogen and progesterone: the two main female hormones, produced by the ovaries in women. Men produce small quantities too.

- Testosterone: Not just a male sex hormone, women also produce small amounts of testosterone, which is an important hormone for sex drive, reproduction and muscle mass.

Women experience fluctuating hormones throughout their lifespan and, with that, come changes in gut function, digestion and gut microbes. We see how our digestion is impacted during menstruation, pregnancy and menopause as I will describe next.

Period poo

Period poos are a real thing and you are not alone in this plight. A change in the consistency, frequency and smell of your poo during your period is very common and, just because we're not all out here exchanging horror stories, doesn't mean it's not happening. You can blame your period poos on hormones.

In case you need a quick biology refresher, the menstrual cycle can be broken down into two stages, the follicular phase and the luteal phase. The follicular phase starts on the first day of your menstrual cycle and ends with ovulation. Ovulation marks the start of the luteal phase and is followed by menstruation. Now, rather than think of the two distinct phases, let's break our cycle into four stages:

- Menstruation
- Follicular phase
- Ovulation
- Luteal phase – then REPEAT....

Digestive symptoms are amplified during the luteal phase and during the first couple of days of your period when oestrogen levels drop. The most common symptoms include abdominal pain, diarrhoea or constipation and bloating. What we know is that oestrogen and progesterone can affect your gut given that they have receptors, or 'docking points', with which they interact along your gut lining. Fluctuating levels can impact how fast or slowly food passes through your digestive tract and the rate at which your gut absorbs liquid.

Add the impact of fluctuating hormones on our mood and you've now got the gut-brain axis involved. Unfortunately, period poos are a very normal and stinky part of life, but what can you do during the luteal phase?

- Avoid caffeine and alcohol
- Adjust your fibre intake depending on the consistency of your poo
- Cut down on fatty and spicy foods

- A low-FODMAP diet may help during this time of the month but only if indicated by your dietitian.

Pregnancy

This chapter in a woman's life brings its fair share of aches and pains, including digestive challenges (because that's exactly what we need as we prepare our bodies to birth humans, NOT). The most common complaints at certain points throughout pregnancy include constipation, reflux and/or heartburn. During pregnancy, progesterone levels remain high, causing food to move much more slowly through your gut subsequently, bringing on the not-so-joyful experience of constipation. That accompanied by straining increases your chance of haemorrhoids (Chapter 11).

Pregnancy hormones are also responsible for symptoms such as nausea, as well as reflux and heartburn. These hormones can cause your LOS (remember the sphincter we've spoken about connecting your oesophagus and stomach?) to relax, in turn causing stomach contents to travel back up your oesophagus. Add the additional pressure of a growing baby, and the chances of heartburn and reflux are uncomfortably high.

Another important change during pregnancy, also due to fluctu-ations in hormone levels, occurs within the maternal microbiome. Research is demonstrating how changes to our inner ecosystem occur more profoundly during the third trimester, when levels of certain hormones are at their highest. For example, high lev-els of progesterone towards the end of pregnancy increase levels of Bifidobacteria species, which can be transferred to the baby during birth and are important for breaking down natural sugars in breastmilk.

Menopause

The chapter in a woman's life in which periods cease, approaching menopause, brings symptoms of hot flushes, sleep disruption and breast soreness as well as gut turbulence including bloating, gas, constipation and reflux. Both oestrogen and progesterone start to fluctuate in the run up to the menopause and then drastically fall, causing your gut motility to slow down, which is why so many women report feeling bloated and clogged up.

My two main recommendations to address a sluggish gut with menopause include:

1. Eating more plants to gradually increase your fibre intake to help with bowel regularity and even positively to influence the gut-hormone axis.

2. Include more soya. Specific compounds in soya (isoflavones) are known to mimic the activities of oestrogen to a certain extent (yet each functions differently in the body so no need to worry about the 'soya and cancer' myth which is a constant floating topic across social media). Think of soya as nature's very own HRT, or hormone replacement therapy, where it can help reduce perimenopausal and menopausal symptoms. Look at incorporating a serving of soya-containing products daily in the form of edamame, soya nuts, tofu, soya milk or soya yoghurt.

The oestrobolome

The connection between gut health and hormones is especially important in the case of the hormone oestrogen. Have you come across the term oestrobolome? If not, you'll be thrilled to know that it refers to a specific collection of gut microbes that are capable of regulating the metabolism of oestrogen. As you no doubt know, oestrogen is one of the most important sex hormones women produce (and men too but to a lesser extent) and plays a crucial part in our menstrual cycle, fertility and pregnancy, bone health, sex drive and heart health. Some important points to be aware of are:

- An imbalance in the levels of oestrogen is linked to an impaired oestrobolome and vice versa.

- Oestrogen comes in contact with the oestrobolome microbe system in the bile.

- The oestrobolome produces an enzyme called beta-glucuronidase, which is essential for the breakdown of complex carbohydrates, the reabsorption of micronutrients such as flavonoids and, most importantly, for the reabsorption of oestrogen and conversion of inactive oestrogen to active forms.

Whether we're trying to support our general wellbeing, our gut health or our oestrobolome in particular, working on the four pillars of health will no doubt offer the benefit we seek. So, without further ado, let's close off the circle and get into the pillars of supporting a blissful gut, starting with nutrition.

Pillar #1: Nutrition

The final 10

As we wrap up all things nutrition, here are my 10 takeaways from this book that I want to make sure to leave you with:

1. Plants should be the heroes of your diet. Whether you are required to follow an elimination diet or not, both I and the science encourage you to aim for 30 plant-based ingredients per week.

2. There is an unfortunate reality when it comes to your wish to 'heal your gut'. It can lead to disordered eating and a turbulent relationship with food. I am seeing this way too often in my practice where I have worked with a great many clients who have been on food restrictions for a long period of time while trying to restore their gut health and have ended up with issues such as:

- Damaging their gut microbial diversity even more.
- Being fearful of reintroducing some foods again such as gluten and a variety of vegetables and fruit.
- Being on the low-FODMAP diet for far too long.
- Developing disordered eating habits.
- Refusing to eat socially.

If this sounds like you or like someone you know, please do not wait any longer to get the support you need, as gut health is all about INCLUSION and not EXCLUSION. If any elimination is necessary, it should be done under the supervision of a qualified dietitian or nutritionist, and for a short period of time until you have identified your exact triggers.

3. Do not take a food intolerance test (unless we're talking about testing for lactose intolerance or fructose malabsorption). I've written about this in detail in Chapter 10 (page 169) but I want to reiterate this warning once again as every client who has struggled with their gut has fallen victim to food intolerance marketing and I urge you not to invest a penny. Remember that these tests are usually very expensive, are unreliable and inaccurate, providing you with a very long list of trigger foods to be avoided so, yep, there we go with EXCLUSION again.

4. Do not take a probiotic supplement unless you have a medically justifiable reason to do so. Bear in mind that you need to know the type of probiotic strain that is required to treat your health condition. It is the strain of the bacteria or yeasts that is most important and determines whether the product is going to work. Instead, shift your thinking towards consuming a variety of prebiotics (page 43). Prebiotics and the more recently recognised postbiotics are currently changing the world of gut health. Prebiotics are, as we've seen, what your gut microbes need to thrive and do their jobs. If I had to pick only eight prebiotics for you to include they would be:

- Jerusalem artichoke
- Onions
- Garlic
- Asparagus
- Raw oats
- Legumes
- Mushrooms
- Cashews.

5. Include some fermented foods in your diet. Fermented foods are foods that are preserved using a process that extends their shelf life and increases their nutritional value while providing some healthful probiotics to your body. Fermented foods contain probiotics that foster the growth of organisms in your gut and help your immune system to work with them to keep you healthy. Incorporating fermented foods into your diet is easy but it is important to note that there are different fermentation methods, some of which do not give you the probiotic benefits. Also, the two most researched fermented foods are yoghurt and kefir. To those who do consume dairy, these tend to be on the list of suggested dairy foods to consume. If you do not consume dairy, fermented cabbage and kombucha are the next two choices to include regularly as part of your diet. All you need are small amounts, consumed regularly.

6. Work on your plate ratio and get your 'gut plate' on. Another way to get more plants into your diet is to work on your gut-plate. Use the plate model shown in Figure 19.

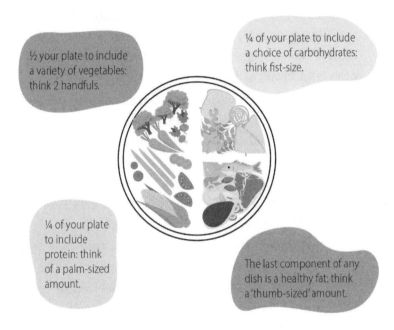

½ your plate to include a variety of vegetables: think 2 handfuls.

¼ of your plate to include a choice of carbohydrates: think fist-size.

¼ of your plate to include protein: think of a palm-sized amount.

The last component of any dish is a healthy fat: think a 'thumb-sized' amount.

Figure 19: The blissful gut plate

So...

Fill ½ your plate with non-starchy vegetables with a minimum of three different types:

- Dark green Swiss chard, bok choy, broccoli
- Spinach, carrots and courgettes/zucchini
- Brussels sprouts, aubergine/egg plant and spinach
- Tomatoes, kale and beetroot
- Lamb's lettuce, grated carrots, cucumbers and cherry tomatoes
- Courgettes/zucchini, cauliflower and peas.

Fill ¼ of your plate with meat or meat-alternatives:

- Palm-sized amount of grilled fish such as salmon or trout, or poultry.

- Eat a variety of plant-based protein-rich foods, such as tofu, legumes, seeds, nuts and nut butters. Examples of legumes include kidney beans, red and green lentils, split peas, chickpeas, black beans, navy beans, and black-eye peas.

- For plant-based recipe ideas, try vegetarian and Mediterranean Diet cookbooks and other cuisines that include plenty of plant foods in their traditional cooking (e.g. Indian, Jamaican, Middle Eastern).

Fill the final ¼ of your plate with high-fibre carbohydrates (i.e. starchy vegetables and grains):

- Starchy vegetables include potatoes, yams, corn, and sweet potatoes.

- When choosing grains, make most of them whole grains. Examples of whole grains include quinoa, oats, whole rye, corn, brown rice, barley, bulgur, spelt, whole wheat and millet.

- Do not demonise 'white-starches' such as rice and pasta as they can all be part of a well-balanced diet.

To complete your meal, don't forget about fats. Think olive oil, nuts or seeds, avocado or even a little dairy, such as yoghurt or feta.

7. Make friends with fibre. As I hope you've taken on board while reading this book, you will need to establish the right amount of fibre that works for your gut. I've presented you with multiple ways to add more fibre into your diet, but the ultimate goal is consistency with diversity. Yes, you will encounter numerous situations where you won't be able to hit 30 plant points per week and that is okay. Establish a strategy that doesn't add additional stress, such as relying on frozen vegetable mixes temporarily during busy times or opting for canned legumes when having to soak and cook dried beans is the last thing on your agenda.

8. No, it is not 'leaky gut'. Throughout your quest to heal your gut, you most definitely will have come across a rabbit hole of information about how certain foods have caused your leaky gut. 'Leaky gut' is a condition characterised by damage to the lining of the gut and it is said to be the cause of numerous gut disorders, including Crohn's disease. Some so-called 'experts' have also claimed that leaky gut is to blame for some autoimmune disorders. Basically, it is used as a 'catch-all', umbrella term for almost every intestinal condition out there.

First, let's start with the term and its definition. 'Leaky gut' is a non-medical marketing term, and it has unfortunately been used to push several 'miracle' cures, fad diets and unhealthy 'quick fixes' to very real gut health concerns. The medical term for leaky gut is 'intestinal permeability'. The lining of our gut (intestine) controls what passes from inside the intestine into our bloodstream; intestinal permeability just means that intestinal barrier function is compromised and large, unwanted particles can 'leak' into the bloodstream. A big claim about leaky gut is that it is the cause of conditions such as IBD, coeliac disease and other gut conditions. However, intestinal permeability can be considered a symptom of these conditions, not the cause. The medical world is indeed researching more deeply into how alterations in intestinal permeability are involved in autoimmune conditions such as type 1 diabetes and coeliac disease as well as gut conditions such as Crohn's disease and IBS, but nothing is conclusive to date. Attempts to improve leaky gut have not resulted in improvement of overall symptoms. There is also no medical backing to support claims that supplements can heal or improve leaky gut. So... is leaky gut real?

Intestinal permeability is very real; however, the claims that it is the cause of a number of conditions and that there are miracle cures out there are not. Before you start removing food groups from your diet and taking unnecessary supplements to treat your leaky gut and cure yourself of every single gut condition known to man, take a step back, be critical of what you read and consult a professional.

The best way to protect your gut is to apply the previous steps I've just mentioned. While it would be incredible for one condition to be the cause of numerous food intolerances and autoimmune disease, it is irresponsible for anyone to claim that is the case.

9. Listen to your gut. If your gut is unhappy, here are some warning signs that something may be up:

- Your skin is acting up
- You experience painful bloating regularly
- Diarrhoea or constipation
- Constant fatigue
- Increased stress or anxiety
- Reflux and/or heartburn.

When you start to encounter your body's warning signals mentioned above, start temporarily to track your food and symptoms. Knowing and keeping track of these symptoms can help you start to identify patterns, and will ultimately help you and your gut health specialist get to the bottom of your gut health issues.

10. Choose your gut health professional wisely. Within the crazy world of wellness and social media, everyone's a gut specialist or an expert now. When it comes to gut health, here are some pointers to keep in mind starting off with the big question: Who is a gut 'specialist'?

- First and foremost, a GASTROENTEROLOGIST
- A gastroenterology-specialist (i.e. gut health dietitian)
- A clinical nutritionist specialising in gut health
- Clinical dietitians and nutritionists who specialise in gut health need to have a minimum of five years' experience working in the area of gastroenterology.

Who is NOT a gut 'specialist':

- A self-proclaimed health guru who acquires their medical knowledge via spirits
- A coach with a three- to six-month online diploma in gut health
- A health coach with no clinical experience
- Your family, your friends, your neighbours
- A food blogger
- A wellness influencer promoting probiotics.

I can safely say that I'm a specialist in gut health after 10 years of working in the area of gastroenterology (both in hospitals and private practice), after working with hundreds of patients and having to complete hours of CPD (continuous professional development). Another crucial warning to bear in mind: health coaches are NOT nutritionists. I want to bring this up due to consulting clients who have been 'diagnosed' by health coaches and prescribed a ridiculously strict diet to 'cure their gut', but instead, have found this has made things much worse. Firstly, health coaches are excellent members of your wellbeing team and I'm lucky to know and work with some great health coaches. What I have seen though is the public assuming that some health coaches have the same role as dietitians or registered nutritionists and, legally, that is not the case.

A health coach is your ally when it comes to achieving your goals. They can help you navigate your health goals by connecting you to the professionals who best fit your case. Health coaches tend to have good knowledge about basic nutrition principles and can encourage clients to eat well, move more and live their best life. What health coaches LEGALLY CANNOT DO is:

- Prescribe diets (medical nutritional therapy) to treat and manage health conditions.

- Treat symptoms of medical conditions.

- Diagnose medical conditions.

- Prescribe supplements (but can offer evidence-based insight about them).

I am seeing lines being crossed, medical conditions being mismanaged and health coaches prescribing diets to manage conditions related to gut health and autoimmune disorders. These actions are ethically and professionally wrong! Having said that though, here's how I work with health coaches:

- A health coach approaches me for dietary advice and management of their client's condition.

- I closely work with the client and the health coach, where their role is to facilitate and help the client implement those guidelines into their daily life.

- I offer contacts to address the 'mind' pillar of gut health – e.g. psychologists – where the health coach can encourage clients to approach, book and attend their sessions.

- We all work as a TEAM to optimise our client's wellbeing and gut health.

Unfortunately, many health coaches have gone rogue, putting their clients' wellbeing at risk and misrepresenting their profession. To end this slight rant, I want to acknowledge all the health coaches

I know who are truly making a difference and continue to respect their scope of practice.

Pillar #2: Mind

Earlier in the book, we talked about the gut-brain axis and the crucial role it plays when it comes to influencing our digestion as well as the health and diversity of our gut microbes.

The reasons why I wanted to elaborate more on the mind pillar are mainly due to recent scientific breakthroughs we are witnessing that show how a dysbiotic inner ecosystem is linked to several mental illnesses, including anxiety and depression. Furthermore, the COVID-19 pandemic has placed an immense strain on our mental wellbeing with reports that mental health problems are at an all-time high. With the gut-brain axis being the bidirectional communication pathway connecting our brain and gut via the vagus nerve, there is no doubt that signals from an anxious or depressed mind will negatively impact the diversity and functioning of our ecosystem and the opposite also applies. If we lack an abundance of helpful gut bacteria that produce compounds which positively impact our mind and mental health, our minds may experience turbulent times given the information being sent from our gut.

The vagus nerve and psychobiotics

Let's look at the vagus nerve (VN) for a minute. The VN acts as the mediator of the gut-brain axis and has a list of life-dependent roles to play such as breathing and regulating our heart rate and immune system, as well as our gut motility, digestion and appetite. From an anatomical perspective, this 'wandering' nerve is actually a network of nerves that are central to our body's autonomic nervous system – the system that controls involuntarily bodily functions, meaning we cannot consciously control them. This system is divided into three distinct networks:

1. The sympathetic nervous system: the network responsible for activating our 'fight or flight' response.

2. The parasympathetic nervous system: the network that helps us to rest, digest, relax and calm down.

3. The enteric nervous system: also known as our 'second' brain, this is a complex network of nerves found in the upper part of our digestive tract and which can function entirely alone, independent of our central nervous system (i.e. our brain and spinal cord).

The VN is a crucial facilitator of the enteric nervous system where our gut microbiome utilises its communication services to modulate our mind. If we had to look at specific gut microbes that communicate with our brain, the emerging field of 'psychobiotics' is opening up a portal into the world of specific probiotic strains and their impact on mental health, specifically anxiety, depression and stress resilience.

Psychobiotics are defined as probiotic bacteria that, when ingested in appropriate quantities, exert a positive health benefit on mental wellbeing including psychiatric illnesses. Despite this definition, these probiotics have not been formally included in the treatment of mental health conditions. The exact mechanisms by which these microbes influence mental health include:

- Reducing the levels of the stress hormone, cortisol and improving the functioning of the VN.

- Producing neurochemicals (i.e. hormones and neurotransmitters) such as serotonin, also known as our happy hormone, and GABA (gamma-aminobutyric acid), the neurochemical responsible for calming our nervous system.

- Reducing inflammation by limiting the production of inflammatory chemicals called cytokines.
- Setting off neural signals along the gut-brain axis.

Given the involvement of psychobiotics in numerous processes connected to our mental wellbeing, they are becoming an exciting avenue for researching treatment options for depression and anxiety. If we were to look at depression specifically, about 300 million people suffer from this condition globally. The connection between depression and gut conditions such as inflammatory bowel disease (IBD) and irritable bowel syndrome (IBS) has been observed and studies have shown that those with depression present with poor microbial diversity and a lack of richness when it comes to specific strains that influence our mood.

From a nutritional perspective and hoping that you've received the memo while reading this book, nutrition is one of the controllable factors that affect our gut microbes, so naturally I want you to think of food as providing ways to keep your bacteria happy, so happy gut = happy mind! This takes us back to the sections in the book that encourage the consumption of fermented foods and a diverse range of prebiotic fibres, and stimulating the production of compounds like butyrate.

The mind toolbox

As we address the mind pillar of gut health, there is enough research out there to encourage you to find ways and activities that maximise positive communication along the gut-brain axis and support the function of your VN. In Chapter 7, there are therapy options for managing IBS by targeting the gut-mind connection, but here I want to add a few more suggestions to your toolbox:

Meditation: Yes, this ancient practice has taken the wellness world by storm but it is definitely not just another trend. Meditation is all about training yourself to be more aware, to be more in tune and to observe your thoughts sans judgement. Consistency is key here. There are different types of meditation out there and I have personally found various forms that I can incorporate into my day, such as guided meditations at the end of the day, meditative running in the forest and breathwork.

Nature: Did you know that being in nature can have a huge impact on our stress levels? I remember reading somewhere how, biologically, we are not made for our current society with constant, daily stressors triggering our 'fight or flight' response. Multiple studies have shown how being in nature can:

- Reduce anxiety and anger
- Reduce muscle tension and heart rate
- Reduce blood pressure
- Increase positive emotions and overall wellbeing.

My personal goal as of 2018 was to be in nature as often as I could and truly disconnect whenever possible. So, whether it's a walk in the forest, by a lake or in a park, try to get out there at least once or twice a week and appreciate how mother nature can heal.

'Brain-dump': This is an activity that can help busy minds before bed. I found it extremely useful to come up with a nighttime ritual that consisted of a cup of chamomile or lemon balm tea and a diary to brain-dump into. The way I would describe brain-dumping is 'word-vomit', or writing down every thought and feeling that comes to mind. Don't worry if whatever you have noted down doesn't make sense. Just keep going until you feel you've 'cleared' your head. Dumping all these thoughts on a piece of paper right before bed has helped me tremendously with my anxiety and sleep.

The power of the breath

I want you to stop for a second and notice your breathing. Are you a shallow breather or a belly breather? That is, are you breathing into your belly or are your inhalations more of a shallower, chest-type of breathing?

There is no way of discussing relaxation techniques without mentioning diaphragmatic breathing or deep 'belly breathing'. Simply put, it involves breathing deeply into your belly where it expands outwards with your inhalation and contracts with your exhalation. This technique aims to slow your respiratory rate, in turn, lowering your pulse rate and blood pressure and quietening your sympathetic nervous system – that is, the 'fight or flight' response. How does this affect your gut?

Diaphragmatic breathing affects the gut-brain axis by engaging your vagus nerve and thereby stimulating the parasympathetic nervous system to 'rest and digest'. Think of it as if you were tapping someone on the shoulder to get their attention. The benefits your gut will experience when practising regular deep-belly breathing include:

- Reducing abdominal pain and the sense of urgency
- Improving digestion
- Expelling gas easily
- Breaking the cycle of setting the internal alarms when your gut experiences diarrhoea – instead accept, reflect and move forward.

Deep belly breathing has been practised for hundreds of years, whether as part of yoga or of other traditional practices such as tai chi. Now, the science is catching up to understand the impact these ancient practices have on our physiology and, not surprisingly, diaphragmatic breathing is proving to be an inexpensive, self-administered intervention that can reduce both physiological and psychological stress. Physiological markers that were measured and improved in some studies included blood pressure, respiration and cortisol levels (a known stress hormone). Nevertheless, we still require ongoing research to uncover more; however, there is absolutely no doubt that deep-belly breathing can be beneficial and safe.

Figure 20 shows a simple exercise I want to share with you and it involves my go-to body check-in when I'm going through an anxious and stressful phase. I tend to check in and wind down with these three steps at the end of the day and/or in between clients and meetings, so here goes:

 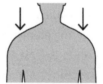 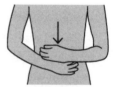

1. Relax your jaws: Notice how stiff they are and, if you're like me when I'm stressed, your jaws are locked and hold so much tension. Roar like a lion and give your cheeks a gentle massage.
2. How are your shoulders? Do they tense up all the way up to your ears? Tension in your neck and shoulders is very common when experiencing stress and anxiety. Relax your shoulders, stretch and find ways to get them low low low.
3. Deep belly-breathing: As I've explained, this is a powerful and underused tool that I encourage everyone with a gut problem (or not!) to practise daily.

Figure 20: Three steps to de-stress

My mind pillar story

I failed my own mental health advice only to build a more resilient mental state of mind and gut.

I want to end this section on the mind pillar by sharing a milestone in my life that changed everything, and it came down to figuring out what the missing piece in my very own gut puzzle was. It is a personal story and a snapshot of a period of my life that I hope will resonate with some of you.

Being a business owner, a mother, a wife and a 'me time' advocate, many could describe me as being an over-thinker, a perfectionist, or someone who dives in head-first and is not afraid to fail. I've always had this insatiable thirst for wanting to achieve more and have struggled with celebrating the now and the present. So, with these traits, I've managed to:

- Achieve two university degrees (well, now three by the time this book comes out)
- Work on three different continents
- Run a hospital dietetics department at 25
- Find love and keep a long-distance relationship going for three years (yes, we clearly made it work, almost 14 years together, married with two cheeky little humans)
- Move to Switzerland, where I've now been based for 10 years
- Open up my practice while I was three months pregnant
- Run a successful and stable business.

The reason I've put the above in bullet-points is to help me realise how much I have to celebrate. See, celebrating my successes is something that I have rarely done in my life; instead I have focused on my to-do list and all the things not yet accomplished. At 25, having so much to prove professionally, I was at the point of burnout, when my mental and physical health took a blow. As you know from earlier chapters, I went through a year of gut hell, eventually being diagnosed with severe IBS, anxiety and stress. My immune system suffered and it was simply a very unpleasant time. My triggers? Being over-worked and under-appreciated, over-worried about my long-distance relationship and over-thinking my future. Talking about emotions, feelings and anxiety back then (and to date) was taboo. I also underestimated the power of the mind and how our mental struggles can manifest into physical symptoms.

What did I do? I took charge of my health, took up meditation and yoga, resigned from my position and decided to move to Switzerland.

Fast-forward to 2018. In brief, my mind went into overdrive and once again, I'd lost the concept of what balance is. Being an over-thinker and over-achiever, my days had no end. I also decided to go back to studying and got accepted into the International Olympic Committee's Sports Nutrition Program. Everyone called me crazy or simply commented, 'I don't know how you do it.' I liked that.

I tend towards perfectionism where I strive to be the best mother, the perfect wife, the successful business owner and the one who can do it all. I became so overly consumed with perfectionism that I no longer realised my red flags — inconsistency with my workouts and meditation, becoming sick quite often, constantly playing catch up with work and studies, having my first panic attack, no longer being a social wild-child (actually, that change comes with parenthood) and not being fully present in the moment with my boy back then. Oh, and my gut literally went to to pot as they say.

My panic attacks became more frequent, and I finally learned that was what they were after jumping from one doctor to another to find answers, including ruling out asthma, and finding it all came down to stress and anxiety as the cause of my symptoms. After that I didn't wait too long until I got myself into therapy, realising that I needed to defer my studies and that I was and am perfectly imperfect. Coming to that realisation was probably one of the hardest things I've had to do but here's how my journey unfolded:

- I started off with weekly therapy sessions and it's now been four years (five by the time you read this book) of consistent monthly sessions.

- I meditated daily (twice a day at times) for two years and have found that I can now pull in and out of meditation as needed.

- Professionally, I focused on my weekly accomplishments and learnt to bucket my work-time (i.e. set a limit on my working hours).

- Now I continue to apply what I preach to my clients in terms of stress and nutrition.

- I have had to go back to low-FODMAP eating and strategic probiotic supplementation to restore my gut health once again, but here I am now, sans eliminations, restrictions and rare flare ups.

- I relearnt how to relax and waste time without feeling guilty.

- I am BREATHING: Breathing through tension, anxiety, stress, sadness, happiness, conflict; I am learning how to breathe through it all.

Being vocal about my struggles and not feeling ashamed about my journey has also given me the strength and confidence to overcome my biggest challenges. Seeking help and realising that something is not quite right is never a sign of weakness. I have mentioned this numerous times throughout this book but do seek the help that you need because it is out there. My missing piece in achieving the coveted blissful gut was the 'mind'. Mental health matters now more than ever, and if your mind is struggling, I am more than certain that your gut will follow suit.

Pillar #3: Sleep

It's 11:45 pm and Jameel is staring into the darkness as his mind races, recapping snippets from work. He turns, reaching out to his partner as she lies there sound asleep and can't help but feel a tinge of jealousy for how quickly and deeply into rest she has fallen. He spends the next hour tossing and turning, wondering whether sleep will ever grace him with its presence. His mind then races through the day's conversations about the uncertainty of his position given the impact the pandemic has had on sales to date. 'We are trying our best to secure more funding for your position,' the start up's founders had uttered. And that was his final thought before drifting off. The alarm is set to go off every morning at 6:20 am. He literally crawls out of bed, longing for a few extra minutes of shut-eye and rushes to get ready, sprinting out of the door to face a new day at work. His multiple cups of rousing coffee fail to provide him with much needed focus and energy and off he goes, running to the porcelain throne, where his gut has decided to fail him yet again, just as it has repeatedly these past couple of weeks.

Jameel (not his real name) was diagnosed with IBS-D a couple of years ago and, for him and the many 'Jameels' who have consulted me in my practice, stress and poor sleep are the perfect brew for an IBS flare up. Sleep deprivation over a prolonged period of time is associated with an increase in stress hormones and inflammatory markers, in turn, negatively impacting our gut. We also know that our gut microbiome has a sleep-wake cycle that matches our own circadian rhythm, so if we've slept poorly, so have they.

In Chapter 7, we spoke about the importance of addressing 'sleep hygiene' to optimise your gut, mood, immunity and overall wellbeing. This advice is applicable to pretty much every gut condition out there, including IBS, and while we are still at the early stages of uncovering the microbiome-sleep relationship, we are seeing glimpses of how this connection translates. Here are the most recent findings:

- A diverse microbiome promotes healthier sleep.
- Specific gut microbes may regulate sleep quality and duration by controlling chemical signals via the gut-brain axis as well as our immune system.
- Some research has looked into 'good sleepers' and how they may have specific microbes (e.g. Verrucomicrobia and Lentisphaerae phyla) that are associated with performing better on cognitive tasks.

- Serotonin, GABA and dopamine all have an important role in influencing sleep and we are learning how our gut microbes are involved in the production of these hormones.

So where does this leave us? While we don't have a probiotic prescription to improve your sleep yet, I just want to use this opportunity to nudge you towards the importance of taking action and addressing your sleep. Let's dissect some of the steps to better sleep that we've gone through, with an initial focus on three nutrition-related behaviours to consider.

Avoid caffeine after midday/12:00 pm

Caffeine belongs to a family of naturally-occurring stimulants found in plants. The major dietary sources include coffee, chocolate, sodas, tea and energy drinks. Caffeine is also found in some non-prescription medications and more recently in sports supplements. Once consumed, it is rapidly absorbed through the digestive tract where peak concentrations are seen in the blood within one hour after ingestion, with half-life ranging from four to six hours. Everyone processes caffeine differently but we do know that it is a drug that generates a wide range of effects confined not only to physiological enhancements such as alertness and better focus, but it can also affect the system as a whole, producing negative side effects such as nausea, anxiety, insomnia, stomach upset, increased heart rate and restlessness.

When it comes to its impact on sleep, caffeine has been found to negatively affect its quality and duration and, given the wide range of its half-life, it is best to keep that last cuppa for the mornings, especially if you struggle with sleep in general. My recommendation is to switch to decaf after midday or, if you do consume a caffeine-containing beverage in the afternoon, avoid consuming it six hours or less before bedtime.

Final meal at least three hours before bedtime

You've definitely heard this at some point if you've asked a nutritionist when is the ideal time for your last meal of the day. Generally speaking, the suggestion to keep your last meal three hours or more before bedtime allows time for digestion and for food to move from your stomach into your lower digestive tract.

Portions are another thing to consider as a large, late dinner has been associated with poor sleep quality. What we also may fail to realise is that food consumption causes the body to produce insulin, which is a hormone that opens up your cell doors for the broken down food (specifically sugars or glucose from carbohydrates) to be used up for energy. Signalling your body to start using up the energy from food can result in wakefulness and impact your circadian rhythm.

Finally, we've learnt that lying down after a meal can worsen reflux and heartburn, so the general rule-of-thumb to go by is be sure to consume your final meal, three hours before bedtime. There are exceptions of course. Those who suffer from type 1 diabetes or any other condition that warrants the consumption a small meal before bed have to follow different rules. Shift-workers will also require a more tailored approach so, when in doubt, always seek support from your qualified dietitian or registered nutritionist.

Trigger foods NOT in the evening

This may seem like common sense but it is something I always remind my clients of when they are embarking on the challenge phase after a period of elimination for a suspected food intolerance. We clearly do not want to stimulate our bowels right before bed as the discomfort of bloating, stomach pain and diarrhoea aren't a particularly good recipe for a restful night's sleep. For this reason, if you are currently reintroducing trigger foods as part of a challenge phase, do so during the day and not at dinner time. Furthermore, do not forget that trigger foods may also include alcohol and spicy foods; if those two are culprits, I would suggest avoiding them at your final meal of the day.

Before we continue talking about more sleep insights, I'm repeating Figure 14 here as a little reminder of your sleep-hygiene guide so that you don't have to go back to Chapter 7, again...

Figure 14 (repeated): Steps to improve sleep

How do you know if your sleep is in a good state?

I would ask yourself this one simple question: 'Do you feel rested when you wake up?' Funnily enough, this one question uncovers whether the quality, quantity or both of sleep need a bit of work. You're doing well with the sleep pillar if:

You answer yes to this question

You wake up without the need of an alarm

You don't rely on caffeine to survive the day

You're not making up for lost sleep over the weekends.

And on the topic of hours, do we all need eight hours of sleep? Research suggests that for optimal health, adults should be aiming for seven to nine hours of sleep regularly. If we get less than seven hours of sleep consistently, we may increase the likelihood of weight gain, poor immune function, high blood pressure, increased pain and impaired performance.

Two common complaints that a lot of you voice include not being able to fall asleep easily or not being able to fall back asleep if you've woken up in the middle of the night. Figure 21 on the next page shows a mini-road map of how to tackle both, highlighting different combinations of strategies mentioned throughout this book.

The one that ends with a blissful gut

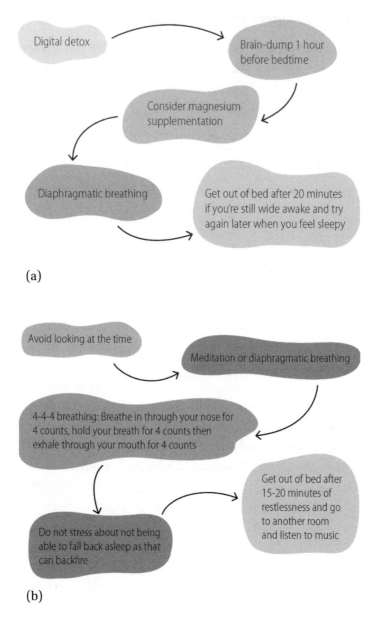

(a)

(b)

Figure 21: How to get to sleep if you (a) can't sleep; (b) can't fall back asleep

Natural sleep aids

Before we get into the topic of sleep aids, I want you to know that you don't have to rely on supplements to solve your sleep woes. Addressing the 'WHY' first is always recommended and, as tempting as it is to ignore the bigger issue and rely on temporary fixes, you and I both know that it won't be sustainable in the long run. It may be hard work finding your way towards better slumber, but it will be worth it.

A response to sleep struggles is to seek help and consider the use of popular sleep aids. Despite being available over-the-counter, I cannot stress enough how important it is to consult your doctor first before considering any owing to the possible side effects. Remember, 'natural' doesn't always imply 'safe'. Moreover, if you have been struggling with sleep for more than two weeks, then it is definitely time to consult a specialist. I've listed the most commonly used sleep aids below, starting off with melatonin.

Melatonin

A hormone that is essential for regulating our sleep-wake cycle, melatonin is mainly produced in the brain by the pineal gland. However, recent studies have demonstrated that melatonin is also produced in the gut, adding more roles to its repertoire that includes involvement in regulating gut motility, its impact on gut microbes and the production of short-chain fatty acids (SCFAs) as well as its important role in regulating our immune system.

To keep the focus on sleep (with a touch of gut talk a few sentences down), melatonin levels naturally increase in the evening and reduce in the morning. This makes it a tempting contender as a sleep aid in the form of supplements, and studies have shown that it may improve overall sleep quality, reduce the time it takes to fall asleep subsequently and increase the duration of sleep. Despite studies showing its safety for adults using it for a short period of time in doses of 3-10 mg, we lack research on their long-term use, including long-term safety and effectiveness. Those who may benefit the most from melatonin supplementation are shift-workers and travellers combating jet lag.

With regard to gut health, melatonin supplementation has recently been shown to alleviate abdominal pain in those with IBS due to its analgesic effects. We do not have enough research to recommend it as part of managing IBS but I thought you'd like to know that the science is pretty much looking at every possible route out there when it comes to treating gut conditions.

Magnesium

We've come across magnesium as a popular relief for constipation given its 'muscle relaxant' effects, but it has also been used as a sleep aid due to its role in regulating melatonin production and its impact on levels of GABA (gamma aminobutyric acid), the calming neurochemical. Symptoms of magnesium deficiency include insomnia and sleep disorders, so no wonder this mineral is a necessity for sleep. Before jumping on the supplementation bandwagon, I'd like to opt for a food-first policy, so sources of magnesium include:

- Wheat bran
- almonds
- Brazil nuts
- cashews
- edamame
- flaxseed
- legumes
- okra
- quinoa
- seafood
- soya products (milk and yoghurt)
- spinach
- sunflower seeds
- Swiss chard
- tamarind
- tempeh
- tofu
- yeast extracts (including Vegemite/Marmite).

Should you wish to trial magnesium supplements for sleep, start at a dosage of 150 mg and do not exceed 350 mg. Before you do, always consult your health professional first to determine the right dosage for you.

L-theanine

This compound, which is naturally found in green tea leaves and some types of mushroom, has been studied as a sleep aid on the basis that it is believed to boost the levels of our brain's calming chemicals, such as GABA and dopamine, at the same time reducing chemicals that are associated with stress and anxiety. In brief, L-theanine fosters a good night's sleep by promoting relaxation and reducing anxiety. Based on the available research, the recommended dosage is anywhere between 100 and 400 mg when used as a sleep aid. As with supplements mentioned above, it is advisable to start at the lowest dose first.

Pillar #4: Movement

There is no doubt that movement in all its forms will offer all-round health benefits, our guts included! We looked at the importance of movement when addressing constipation and as part of living with IBS, but let's go a little deeper to see where the research currently stands when it comes to movement for a blissful gut.

Generally speaking, any form of movement is going to be good for your gut but the science has mainly focused on cardio-based activity such as cycling, running and gym-based workouts. There are three noticeable benefits based on the current research:

- Exercise boosts the number of beneficial microbial species.
- Regular movement enriches microbial diversity.

- Exercise improves the development of 'commensal' bacteria – the microbes that interact with our immune system to enhance protection against pathogenic invaders.

Now, if you're thinking you should ramp up your current efforts with movement, don't undermine the power of consistent low-intensity exercise. Regular, low-intensity movement, such as walking, swimming, Pilates and gentle spinning, has been shown to have protective effects against colon cancer, diverticular disease and inflammatory bowel disease. Why? This is mainly due to regular movement allowing less time for stool to sit in your gut, exposing its lining to potential pathogens. Simply put, it keeps the traffic moving!

Is exercise your nemesis?

More often than not, people have been programmed to see exercise as something that they are 'supposed' to do, which I believe serves as a recipe for inconsistency and ultimate failure. Perhaps we should change our views as to why we exercise in the first place. My story with exercise started after my first year of university, when I piled on 15 kg thanks to a diet of alcohol, freshly baked muffins, fast food after hours and well, first year 'uni-life' really (shocking for a first year dietetic student!). As I slowly started to improve my habits, I decided to enrol in a gym for the very first time. The journey to shift that number on the scale started and 10 months later, I had managed to lose 13 kg and develop an obsession with 'working out' to keep that number from going on an upward trend again.

As an 18-year-old, you lack a true understanding of the physiology behind weight fluctuations, with an expectation that you will remain at the same weight throughout your entire life. So, with any slight increase, I was back at the gym, breaking into a sweat, paying the price for any indulgence. I firmly believe that the main cause of my struggles with exercise and consistency stemmed from that experience. And so, my wiser-self has a few tips and tricks if you're struggling:

1. Start addressing the reasons behind why you would want to start moving and get more active. Forget about weight loss and focus more on strength and enabling your body to perform better.

2. Being active comes in many forms so find something that you enjoy doing as you're more likely to keep that up long-term. With my family, we spend so much more time outdoors than we did in previous years, whether it is for our weekly forest walks or alpine hikes. Start a weekly routine with your family that involves some sort of activity.

3. Avoid gym memberships unless you plan to visit AT LEAST three times per week and you enjoy doing so.

4. Commit to a routine. I found that blocking off my calendar two days a week just to train made it as important as any other work meeting. Time is really no excuse.

5. Invest in a personal trainer if that is the only way to get you moving. Personally, I need movement accountability. Since I started with my personal trainer postpartum, with a focus on recovery and strength training, I can honestly say that this has been the best investment in my road to self-care. The 'improved body composition' was actually a side effect and was never a focus but I've learned that I need to be challenged and guided.

6. See movement and training as a show of self-respect rather than punishment. Don't look at exercise as a means of compensating for indulgences because that simply pushes you into a vicious cycle of dieting and guilt. Respecting our bodies involves nourishment and moving – moving for strength, agility, injury prevention, regular poos, better gut function and flexibility.

It took me years to figure out how I could keep 'movement' consistent as part of my life, so take the time to understand your approach and attitude towards 'movement' before mindlessly deciding to start an exercise routine, which may or may not work for you.

The WYP (walking, yoga and pelvic floor exercises)

If we had to look at just three exercises to incorporate for better gut health, I would suggest starting off with what I'd like to call the WYP.

Walking is inexpensive, requires no equipment except for a good pair of shoes and is perfect for those starting on their movement journey. If you're able to, walking activates your digestive tract to move and contract and, with waste moving through your intestines, your bowel movements remain regular with less bloating. Add the element of nature – say, walking through a forest, by a lake or along a beach – and it can be the easiest stress-reliever there is, addressing both the mind and movement pillars at the same time.

Yoga may sound a little 'woo-woo' to many but it has accumulated enough science to show that this traditional practice can improve gut symptoms and reduce inflammation. Yoga is a centuries-old 'mind-body-breath' discipline that targets the gut-brain axis. Breathwork that is taught in yoga activates our parasympathetic nervous system (aka the rest-and-digest system) aiding relaxation but also helping many cope with painful stomach cramps or other uncomfortable gut symptoms. The yoga moves or 'asanas' involve a number of postures and sequences that can improve digestion, offer your intestines the twists and stretches to soften any tension and release any uncomfortable trapped gas, making you toot better. Research has even shown that yoga can be as effective as a low-FOD-MAP diet for reducing symptoms in those diagnosed with IBS. Furthermore, one study showed how an hour of yoga a day reduced symptoms associated with inflammatory bowel disease (IBD).

Pelvic floor exercises to strengthen that part of your anatomy are important for maintaining urine and bowel continence. You can think of your pelvic floor as a group of muscles that form a trampoline, extending from your tailbone to your pubic bone and from one side of your sitting bones to the other. Pelvic floor muscles help with peeing, pooing and having sex and are known to weaken as we age and due to injury and childbirth. For this reason, don't wait until you have a medical issue that forces you to start thinking about your pelvic floor. Start taking care of that part of your body and activate your pelvic floor muscles to ensure you maintain excellent bladder and bowel function. This three-step exercise is a good place to start:

1. Whether lying down or seated, squeeze and draw the muscles around your anus as if you are lifting them upward for 8 seconds. For the ladies, just imagine that you are trying to draw in a tampon with those muscles without squeezing your bottom. Remember to breathe slowly.

2. Relax your pelvic floor muscles for 8 seconds.

3. Repeat this sequence three times a day, whether you're at your office desk or even when brushing your teeth.

Runner's gut

If movement is essential for gut health, why can it trigger so many digestive issues, especially in recreational and professional athletes? Enter the phenomenon called 'runner's gut'...

Have you experienced that urgency to find a bathroom out of the blue, mid-run? You're not alone! 'Runner's gut' is real and can be disruptive. It can ruin a perfectly good training session but it can also create a lot of anxiety around training which can compromise performance. What actually triggers it?

- Reduced blood flow to the gut: This is the main cause of abdominal pain, vomiting and diarrhoea. During exercise, blood is redirected away from the gut by 20% and up to 80% after one hour, which can create issues.

- Dehydration: It can trigger nausea, vomiting and other gastrointestinal problems during exercise. Always start your training sessions well hydrated.

- Timing of the last meal before exercise: Try not to eat too close to the time you'll be exercising. Avoid eating a main meal within the two to four hours before exercise.

- The type of food eaten just before exercise: Meals high in fibre, fat and protein consumed pre-exercise have been shown to cause an increase in symptoms. For some, a modified low-FODMAP meal prior to exercise may offer benefits and reduce any gut discomfort such as bloating and diarrhoea.

The future

Frankly speaking, I am unsure as to how I would like to end this book. Gut health has become an endless tunnel of discoveries and I am fairly certain that, by the time this book has launched, more breakthroughs are going to have been made. For example, in the last few days, I have personally connected with women in science who are doing incredible work in the field of probiotics and colorectal cancer as well as nutritional psychiatry and psychobiotics.

Figure 22: Images of the future

What has enabled these discoveries are the tools currently being made available to us, such as microbiome testing, but once again, hold your horses and don't jump on the first test you can get your hands on. I have briefly spoken about how microbiome testing and analysis is still at an early stage for commercial use despite being marketed everywhere. Scientists have not been able to establish a gut microbiome profile commonly accepted as 'healthy' or 'normal' and that could reliably predict whether a person will develop a particular disease. For curiosity purposes and only if you have the spare money, then you could give these tests a go but only with reputable and leading companies in this field. Most importantly, be mindful of how you interpret any results because if test results indicate that you are missing some 'beneficial' microbes, you may already have ones that we haven't even identified yet that are doing a similar job in keeping you well and healthy. We just haven't named them yet.

By saying that though, our poo has literally opened up a portal to a world that may be central to health and wellbeing.

The headlines will continue to feature work around our gut microbiome, which may well set the scene for what the future holds for our health, possibly powered by our gut. The questions we're now asking ourselves are:

- Will we be able to formulate specific bacterial concoctions for us to take to manage disease?

- Will we be able to advise people specifically on exactly what bacterial types they need to boost and what foods these microbes need to thrive to protect them from disease?

- Could probiotic supplements serve as adjunct therapy to antidepressants?

- Could changing our diet to boost our microbial diversity reduce our risk of Alzheimer's?

We don't have all the answers just yet but I do foresee that the next five to 10 years will uncover more innovative approaches to managing disease using our gut as a centre-point. So for now, I want you to use the power of imagination to construct a futuristic world of personalised nutrition. I want you to imagine this as a world where 'Uber Eats' has upgraded to the 'Uber ME' delivery service. This is a service that sends you a poo test to analyse your gut microbiome and understand your gut woes; then, based on the results, creates and delivers tailor-made meals straight to your doorstep.

You can also get hold of your own 'gardening for dummies' manual on how to care for your internal garden, highlighting how some foods, such as cooked and cooled potatoes, lentils, oats and mushrooms, can help your own garden to thrive. Down the line, this level of personalised nutrition may end our endless pursuit of health and wellness. So, my friends, the future of poo has never been brighter.

Chapter 12

May you no longer have to endlessly search the internet in secret for solutions to your gut woes and I hope that these chronicles have delivered some sort of self-help confidence in gaining control over your gut.

Recipes

Given that I am no award-winning chef but a busy business owner and mum of two, I like to keep things uncomplicated in the kitchen. This small recipe collection that ends the book is just a glimpse into the typical dishes I would make at home (depending on time of course), with the sort of descriptions I would give you if I were to tell you about them over the phone. This section is just a little add-on as I didn't want to turn this into a cookbook; that's another future project perhaps?

The ingredients have been modified to suit different conditions mentioned throughout this book but my number one tip to anyone following a recipe is 'make it your own!'. If a certain ingredient is not within reach or does not work for you, find the next best alternative.

My recipe contributors and Nutrition A-Z team members, Henriette Saevil and Aikaterini 'Katia' Tassiou, have put together their creations in two separate sections so you'll be visiting their kitchens too.

While we try to avoid labelling how anyone eats, these recipes are predominantly plant-based. Keep in mind that plant-based does not equal vegan and for those who do consume animal products, you can still do so and achieve good gut health. It all comes down to frequency and quality.

For general gut health

The following recipes are simply a nod to all things plant, colour and prebiotic. They are also child-approved after passing the taste test sans gagging and tearful refusals. Mind you, I am fully aware of how picky children are when it comes to food and I am very grateful that mine have a palette that has been exposed to spices, herbs and lots of fibrous foods from a young age. Phoenix had his first curry at one and so did Zoe. Now, of course, they both go through phases of refusing to eat certain ingredients, but a quick tip is repeated exposure in different ways.

Quinoa power salad

Ingredients (serves 4)

170 g quinoa, rinsed and drained
30 ml extra virgin olive oil
3 tbsp lime juice
1 tsp lemon zest
½ clove garlic, finely diced
75 g finely diced walnuts and parsley (use a food processor combining about 75 g walnuts and a bunch of parsley, but quantity as you please)
400 g can borlotti beans, rinsed and drained
1 medium red pepper (or capsicum, as I like to call it)
1 medium Lebanese cucumber, finely chopped
1 medium purple haze carrot (or any other carrot), sliced diagonally
10 cherry tomatoes, cut in halves
10 yellow Datterino tomatoes, cut in halves

75 g feta, crumbed

70 g pomegranate pearls

3 tbsp all-seed mix (sunflower, pumpkin, dried soya bean kernels)

Salt and pepper

Optional: Other excellent additions to this salad include olives and baby spinach.

Method

1. Place the quinoa along with 2 cups of water in a medium sauce-pan (optional: add 1 tsp vegetable stock or ½ a cube). Cover and bring to the boil, then reduce to the lowest heat setting. Simmer until the water has been completely absorbed and the quinoa is fluffy. This takes about 20 minutes. (Note: If you do not use all the cooked quinoa for the salad, store the rest in the refrigerator to use the next day.)

2. In a large bowl, whisk together the olive oil, lime juice, lemon zest and garlic.

3. Once the quinoa is cooked, add to a bowl, mixing in the walnut-parsley mixture.

4. Then, add the remaining ingredients: beans, cucumbers, tomatoes, carrots and capsicum.

5. Sprinkle the crumbled feta on top and add the seed mix and pomegranate pearls as the final garnish. Your salad is now ready to be served!

Mushroom and sauerkraut pockets

Ingredients (makes 24)

1 tbsp extra virgin olive oil

70 g spring onion (one small), diced

1 tbsp dried thyme

210 g brown button mushrooms, finely chopped or diced
   (using a food processor)

6 tortilla wraps, cut into quarters (resembles a triangle of some
   sort)

250 g packet of store-bought cooked sauerkraut

75 g vegan cream cheese (use Simply V, natural)

45 g grated parmesan cheese

2 tbsp olive oil or melt 1 tbsp vegan butter to brush the pockets

Salt and cracked pepper to taste

Optional: black sesame seeds or poppy seeds as garnish

Method

To make the filling:

1. Heat the oil in a non-stick pan then add the onion and thyme. Cook until golden brown, then add the mushrooms and cook well.
2. Once the mushrooms and onion are soft and golden brown, add the sauerkraut and cook until the mixture is uniformly combined. Season with salt and cracked pepper.
3. Set the mixture aside to cool down for about 10 minutes.
4. Preheat the oven to 180°C (fan-assisted) and line a baking/oven tray with a baking sheet.

To make the pockets:

1. Grab a tortilla quarter and spread a little cream cheese as a base. Add about a heaped tablespoon of the mushroom filling in the centre.

2. Grate on some parmesan cheese or, if grated already, just sprinkle some on the top.

3. Fold the pockets by first folding the pointiest edge of the tortilla to cover the filling and then you will nicely wrap your mixture by bringing the two outer edges together.

4. Repeat this process until you've lined up 24 pockets onto your baking tray. Brush them with oil and gently sprinkle some seeds on the top.

5. Bake for 10 minutes, until golden brown.

6. Serve as a starter or as part of a sharing snack platter.

Cashew cacao bliss balls

Ingredients (makes 20, depending on how large you roll them)

170 g cashews
125 g pitted dates
25 g almond meal
10 g raw cacao powder
2 tbsp maple syrup (26 g)
1 tbsp coconut oil
2 tsp chia seeds
60 ml warm water
45 g organic peanut butter

Method

1. In a food processor, add the dates, cashews, almond meal, cacao, syrup and oil, and pulse until well combined.
2. Add the chia seeds, water and nut butter to the mix and pulse again to combine well.
3. Once the mixture is well-combined, roll into 1-inch balls (approx. 30g) then place these into an airtight container and store in the fridge overnight for consumption the next day. (You can always store them in the freezer for a later time.)

For plant-based ideas

As much as I loathe labelling the way we eat, I would say the majority of my family's diet at home is plant-centred and I do encourage my clients to showcase plants but without any radical eliminations if they wish still to consume animal products. The following recipes are frequent dishes on our table, including a cookie recipe that features beans.

Bean cookies

Ingredients (makes 12)

400 g can red kidney beans
45 g palm-free, smooth peanut butter
20 g cocoa powder
40 g almond meal
40 g oats
125 ml maple syrup
1 small ripe banana
1 tsp vanilla extract
½ tsp baking powder
Egg replacement (1 tbsp chia seeds, mixed with 3 tbsp water; wait for 5 minutes until a gel forms and use as your replacement)
20 g of dark chocolate choc chips

Method

1. Preheat the oven to 180°C/350°F (fan-assisted) then line a baking/oven tray with a baking sheet.

2. In a food processor, add all of the ingredients listed, except for the chocolate chips. Pulse until a smooth, pasty mixture is formed.
3. Add the chocolate chips and stir to combine.
4. Using a tablespoon, scoop out into 12 portions and slightly flatten with the back of your spoon.
5. Bake for 12-15 minutes until the cookies rise and cracks are formed on the top.
6. Allow to cool for 10 minutes before consumption.
7. Store in an airtight container in the fridge for 2-3 days.

Courgette and feta fritters

Ingredients (makes 12-14 fritters)

For the fritters:
380 g courgette/zucchini (2 medium courgettes)
50 g carrot (1 small carrot), grated
90 g canned corn
1 spring onion, finely chopped
2 tbsp chopped fresh dill
½ tsp baking powder
45 g breadcrumbs
50 g buckwheat flour
2 eggs
65 g feta cheese
3 tbsp extra virgin olive oil
Salt and pepper to taste

For the optional yoghurt-dill dip:
1 small pot (150 g) natural yoghurt of choice
1 tbsp chopped fresh dill
1 small cucumber, finely diced
Salt and pepper to taste

Method for the dip

In a small bowl, add all the ingredients and mix well until you have a creamy, well combined dip. Nothing complicated here!

Method for the fritters

1. Grate the courgettes into a bowl. To get rid of all the excess liquid, place what you have grated on a clean kitchen towel, gather the ends of the towel and twist to wring the excess liquid out. Then separate the grated courgette with your fingers to avoid clumping and transfer back to the bowl.
2. To the bowl, add the carrot, corn, spring onion, dill and baking powder as well as the breadcrumbs and buckwheat flour.
3. Beat the eggs in a separate bowl and then add the mixture to the main bowl.
4. Add the feta and then stir the mixture until well combined to form your 'fritter batter'.
5. Place a non-stick frying pan over medium heat, and add the olive oil. Once your pan is hot, add about 1 heaped tablespoon-sized portion of the fritter mix and use the back of the spoon to gently press the batter down to form a nice, roundish fritter.
6. Pan-fry each side for 2-3 minutes until golden brown and transfer to a plate lined with a paper towel. Repeat the same with the remaining batter.
7. Serve with the yoghurt dip and enjoy.

Chickpea, sweet potato and spinach curry

Ingredients (serves 3)

For the spice mix, combine in a small bowl:
1 tbsp mild curry powder (I use Madras)
1 tsp ground cumin
1 tsp ground coriander
¼ tsp turmeric

For the curry:
1 tbsp olive oil
2 small garlic cloves minced
40 g onion (1 small), chopped
400 g sweet potatoes, cut into chunks
1 can (400 g) chickpeas
Salt to taste
110 g tomato paste (passata)
120 g baby spinach, rinsed
500 ml coconut milk
125 ml water
120 g baby spinach, rinsed

Method

1. In a medium saucepan, heat the oil on medium heat, then add the minced garlic and onion and cook for a few minutes until golden.
2. Add the potatoes and chickpeas and stir thoroughly.
3. Add the spice mix and salt making sure everything in the saucepan is coated nicely.
4. Stir for a few minutes before adding the tomato paste. Combine well then reduce the heat.

5. Add the coconut milk and water, stirring well while you do so. Let your curry simmer for about 20 minutes on low heat until the potatoes are cooked.
6. Add the spinach and cook for a few minutes until it is all wilted.
7. You can serve your curry with a refreshing cucumber salad for a low-carb version or with brown rice.

Reflux-friendly meal inspiration

Reflux and heartburn can definitely take the joy out of eating but you'll find some excitements below that do not include any dietary triggers. Keep in mind that eating behaviours are essential too when it comes to controlling your symptoms, so eat slowly, stop when you are satisfied and not overly full and perhaps drink in-between meals rather than with food.

Ginger-pear overnight oats

Ingredients

For the oats (serves 1):
45 g rolled oats
125 ml unsweetened plant-based milk of choice
50 g natural yoghurt (for this recipe, I use almond-based yoghurt)
1 flat tbsp chia seeds
1 tbsp maple syrup
Toppings: Chopped walnuts or other nut of choice

For the pears (serves 4):
4 pears (make a batch)
1 flat tbsp ground ginger powder
2 tbsp water
1 tbsp maple syrup

Method for the pears

1. Preheat the oven to 200°C/390°F (fan-assisted).
2. Cut the pears in quarters (lengthwise) or in any form you wish.
3. Mix the ginger, water and maple syrup in a small bowl with a whisk, then add to the pears. Make sure they are all nicely and evenly coated with the mix.
4. Bake for 30 minutes or until the pears are soft and golden brown. (Make a batch – clearly I won't ask you to bake just the one pear. I always like to make a batch of baked fruit to keep in the fridge and add as a topping to yoghurt or granola.)
5. Cool and store in the fridge for later.

Method for the oats

1. Add milk, chia seeds and maple syrup to a bowl or jar with a lid and mix well to combine all the ingredients.
2. Add the oats and make sure your mixture is well combined.
3. Press down and make sure all your oats have been immersed in the milk.
4. Place the lid on and let your oats set in the refrigerator overnight.
5. In the morning, enjoy your oats topped with the chopped baked pears and walnuts.

Mushroom and courgette tart

Ingredients (serves 4)

32 cm spelt ready-to-roll dough
140 g vegan cream cheese of choice
1 tbsp extra virgin olive oil
35 g brown onion, sliced (½ a large onion)
180 g chestnut, brown mushrooms, sliced
1 tbsp dried thyme
70 g/1 small courgette/zucchini, sliced into discs
1 egg, beaten

Method

1. Preheat oven to 180°C/350°F (fan-assisted) then line a circular baking tray with baking paper.
2. Roll out the dough onto the paper and prick with a fork, making little holes throughout. Then, evenly spread the cream cheese on the dough.
3. Place a pan over a medium heat and add the olive oil. Once warm, add the onion and mushrooms and stir until they start to soften. Add the dried thyme and continue to cook until the mixture is caramelised and brown.
4. Set the mushroom mixture aside to cool down for a bit. Then, spread it evenly over the cream cheese and dough.
5. Lay the courgette discs over the mushrooms in overlapping circles until you've covered the whole tart. Then, fold the edges of the tart in any form that pleases you.
6. Brush the edges of the tart and the courgette discs with the beaten egg, and bake for 25-30 minutes, until the dough becomes golden brown.
7. Cut in squares or a form you desire, and enjoy with a side salad.

All-the-goodness bowl

This is more of a 'put-together-what-we've made' kind of bowl. Batch-cooking different elements and ingredients can save you a lot of time. The two components I personally like to make and keep in the fridge to incorporate into other meals are:
- Roasted vegetables
- Grain and legume mix.

Ingredients (makes 4 bowls)

For the roast vegetables (I've chosen the following reflux-friendly vegetables):
550 g beetroot (3 medium), peeled, washed and cut into quarters
195 g carrot (2 medium), cut into discs
390 g courgette/zucchini (1 large), cut in half lengthwise then into slices
1 tbsp dried thyme
2 fresh rosemary stems
2 tbsp extra virgin olive oil

For the brown rice and lentil mix:
150 g brown rice (use parboiled brown rice as it's quicker)
140 g beluga lentils

Additional ingredients for the bowl:
250 g purple cabbage, finely chopped
150 g kale, finely chopped
Note: I used a food processor to chop and mix the cabbage and kale.

Ingredients for the reflux-friendly golden creamy dressing

140 g plant-based yoghurt (I use cashew-based yoghurt)

1 tbsp maple syrup

1 tsp Dijon mustard (mild)

1 tsp ground turmeric

1 tbsp oil

1 tsp vinegar

Method for the roast vegetables

1. Preheat oven to 190°C/375°F (fan-forced).
2. In a baking dish, add the vegetables, herbs and oil and combine well.
3. Bake in the oven for 45-60 minutes until the vegetables are nicely roasted.

Method for the rice and lentil mix

1. Cook the rice and lentils (separately) according to the package instructions.
2. Once both are cooked (15-20 minutes for the rice and 25-30 minutes for the lentils), add the lentils to the rice and mix to create your grain and legume mix. (You may have leftovers with the amounts mentioned above but the mix would keep in the fridge for 4-5 days.)

Method for the dressing

1. To a bowl, add the yoghurt, syrup, mustard and turmeric.
2. Then whisk in the rest of the ingredients until you end up with a creamy dressing.

Method to assemble

1. Portion out 120 g of the rice and lentil mix into your bowl as your base.
2. Then, add the different components: roasted vegetables and the cabbage and kale salad.
3. Drizzle with the golden creamy salad dressing and enjoy.

For the faecally challenged

We all get clogged up every now and then and these recipes offer your gut the fibre it needs to get things going. You can always visit my six-step constipation toolbox when in doubt and, remember, if you increase your fibre intake, always make sure you boost your fluid consumption too.

Banana and fig breakfast muffins

Ingredients (makes 6)

Dry
80 g rolled oats
105 g buckwheat flour
40 g walnuts
1 tbsp ground linseed (flaxseed)
65 g dried figs (4 dried figs)
1 tsp baking powder
½ tsp ground cinnamon
¼ tsp ground nutmeg

Wet

1 ripe banana

2 tbsp maple syrup

125 ml plant-based milk

30 ml extra virgin olive oil

2 eggs

Method

1. Preheat the oven to 180°C/350°F (fan-assisted) and grease a muffin tin.
2. Using a food processor, first combine all the dry ingredients, pulsing until the nuts and dried figs are nicely chopped up.
3. Add the wet ingredients and pulse well until the mixture is thoroughly combined.
4. Spoon the batter into the tin and place it in the oven to bake for 25-30 minutes (or when a toothpick inserted in the middle comes out clean).
5. Once the muffins are baked, let them cool on a rack for 10 minutes before popping them out of the tin.
6. Happy pooing!

Pearl couscous and lentil salad

Ingredients (serves 4-6)

For the salad:
230 g whole wheat pearl couscous
165 g brown lentils (washed and drained)
1 medium cucumber (about 210 g), chopped
365 g cherry and Datterino tomato, mixed, cut in halves
65 g pitted dates, finely chopped
2 stems fresh parsley, finely chopped

For the dressing:
35 g tahini
Juice of half a lemon
1 tbsp olive oil
1 tsp maple syrup
80 ml water
Salt and pepper to taste

Method

1. Cook the couscous and lentils (separately) as per your package instructions. Generally, the ratio for the pearl couscous is about 1.5 cups of water to 1 cup of pearl couscous, cooked using the absorption method for about 15-20 minutes. For the brown lentils, the ratio is generally 2 cups of water to 1 cup of lentils, cooking for about 25-30 minutes or until the lentils are soft.

2. Once the lentils and couscous are cooked, set them aside to cool and start assembling your salad.

3. In a bowl, add all the salad ingredients: cucumber, tomatoes, dates and parsley. You can then add the couscous and lentils and mix well.
4. In a separate bowl, place the tahini, lemon juice, olive oil, maple syrup and water and whisk well until you get a creamy salad-dressing consistency. Add salt and pepper to taste.
5. Add the dressing to your salad and combine well.
6. Serve in bowls and enjoy!
7. You can include these optional toppings, which go very well: grilled halloumi cheese or crumbled feta.

The soup and salad combo

This tends to be a recurring combo in our household (we've trained our guts for a ton of fibre over the years) and can definitely aid the faecally challenged too. The soup is a version of my mum's lentil soup, which we always combine with lots of greens – enter this super-simple green salad.

Super-green simple salad

Ingredients (serves 4)

For the salad:
1 large cucumber (200 g), cut into discs
100 g rocket (arugula), washed and dried
140 g baby spinach, washed and dried
230 g corn, canned or in a jar, drained
210 g avocado (1 medium), sliced
65 g nut and seed mix of choice

For the dressing:
1 tbsp dried mixed herbs
2 tbsp white wine vinegar
1.5 tbsp olive oil
1 clove garlic, minced
Juice of half a lemon
Salt and pepper to taste

Method

1. Mix the ingredients well in a bowl or put in a screw-top jar and shake.
2. To a big bowl, add all the ingredients (except for the avocado and nuts); then add the dressing and mix well.
3. Transfer to your serving platter then place the sliced avocado on top and sprinkle with the nut and seed mix.

Mum's lentil soup

Ingredients (serves 4)

1 tbsp extra virgin olive oil
40 g onion (1 small), roughly chopped
1 bay leaf
1 tsp ground cumin
1 tsp ground coriander
200 g potatoes (2 small) peeled, washed and chopped
1 carrot, cut halfway lengthwise, then finely chopped
1 cup red lentils
1 tbsp vegetable stock powder
750 ml water
1 cardamom pod
130 g courgette/zucchini (1 small), cut into small quarters
Juice of half a lemon
Salt and pepper to taste

Method

1. Heat the oil in a medium-sized pot, then add the onion, bay leaf, cumin and coriander and stir.
2. Add the potatoes, carrot and lentils and stir until the potatoes are slightly browned and the vegetables are coated with the spices.
3. Add the stock powder then water and bring to the boil. Add the cardamom pod, then lower the heat and simmer for about 20 minutes or until the lentils and potatoes are softened.
4. Add the courgette and simmer for another 8 minutes.
5. Once the vegetables are cooked, use an emulsion-stick blender to purée into a soup. If the consistency is still too thick for your liking, add some warm water.
6. Add the lemon juice, salt and pepper and serve with the salad.

For bumfluenza

After a case of bumfluenza, your appetite may not be the greatest and the focus should be on rehydration and gradually gaining your strength and energy through food as your appetite returns. The following recipe suggestions are best kept for when you are ready to introduce food, keeping insoluble fibre to a minimum.

Tummy-friendly rice pudding

Ingredients (serves 4)

1 litre plant-based milk (soya, almond or oat)
100 g white rice
1 tbsp sugar
1 cinnamon stick
2 tsp vanilla essence
Toppings: Any fruit of choice that you're able to tolerate.

Method

1. Pour the milk into a medium-sized pot and bring to an intense simmer, which is just under boiling point, and then reduce the heat, add the rice and simmer over a low heat to continue cooking.
2. Add the sugar and the cinnamon stick and stir frequently for about 35-40 minutes until the rice is tender.
3. The mixture should resemble the consistency of runny porridge but will continue to thicken as it cools down.
4. Once cooked, remove the cinnamon stick and then transfer the rice pudding to a large serving dish.

5. Allow the pudding to cool down to room temperature and then refrigerate until you are ready to serve.

Lemon and green pea orzo (Risoni)

Ingredients (serves 4)

1 tbsp extra virgin olive oil
200 g orzo pasta
1 tbsp dried thyme
1 small garlic clove, minced
500 ml vegetable broth/stock
125 ml oat or soya cream
Juice and zest of 1 small lemon
200 g frozen green peas, thawed under warm water
Topping: grated parmesan cheese

Method

1. In a pan, heat the olive oil then add the orzo and stir for about 1 minute until slightly toasted.
2. Add the thyme and garlic and stir until fragrant and well combined.
3. Add the stock, one ladle at a time, stirring until you've added it all. Bring to the boil then gently simmer, covered, for 10 minutes, until the pasta has absorbed the liquid.
4. Remove the lid, then add the cream, lemon juice and zest and peas. Stir until the peas are cooked and a bright green. (Note: try not to overcook the peas.)
5. Season with salt and pepper and, as an optional addition, grate some parmesan cheese on top and serve.

Pumpkin polenta with balsamic roasted beetroot
By Nutrition A-Z's dietitian, 'Katia' Tassiou

Ingredients (serves 4)

For the beetroot:

¼ cup (60 ml) olive oil

3 tbsp balsamic vinegar

1 tbsp maple syrup

½ tsp fine sea salt

¼ tsp black pepper

3 medium-sized red beets, peeled, scrubbed and chopped into small pieces

For the polenta:

3 tbsp olive oil

3 medium-sized shallots, minced

2 garlic cloves, minced

1 litre low-sodium vegetable broth

160 g polenta

250 g canned pumpkin

¼ tsp dried thyme

¼ tsp dried sage

½ tsp paprika

1 tsp fine sea salt

1 tsp black pepper

1 tbsp fresh thyme (optional garnish)

Method

1. Preheat the oven to 200°C/390°F (fan-assisted) and line a baking tin with baking paper.

2. In a small bowl, whisk together the olive oil, balsamic vinegar, maple syrup, salt and pepper.
3. Place the beetroot in the baking tin and then pour the balsamic mixture on top. Toss them lightly with your hands to coat and roast for 15 minutes.
4. Remove the tin from the oven and flip the beets.
5. Return the tin to the oven and roast for another 15 minutes. When the beetroot pieces are done, set them aside until plating.
6. Now, moving on to the polenta, heat a large pan over a low heat. Add in the olive oil, shallots and garlic and sweat for 2 minutes.
7. Turn the heat to medium, add in the broth and bring to a simmer. Then add in the polenta 1/3 cup at a time, stirring constantly to avoid clumps.
8. Reduce the heat to a simmer and cook uncovered for about 20 minutes while stirring frequently until the water is absorbed and the polenta is soft.
9. Lower the heat on the polenta and then add the pumpkin, thyme, sage, paprika, salt and pepper, and then stir.
10. Continue to cook the polenta while stirring occasionally for 5 minutes.
11. To plate, serve the polenta in 4 bowls, topping each with an equal portion of the balsamic beetroot, then garnish with fresh thyme and black pepper if desired and serve warm.

Low-FODMAP (bloat-friendly)

Low FODMAP recipes are in abundance on the internet so I am fairly certain you won't be short of ideas. These recipes are applicable whether you are going through a period of uncomfortable bloat or are strictly following the low-FODMAP diet.

Tofu soba noodles bowl with peanut sauce
By Nutrition A-Z's clinical nutritionist, *Henriette Saevil*

Ingredients (serves 2)

For the noodle bowl:
100 g soba noodles
1 tsp sesame oil
220 g tofu cut in dices
100 g carrots, thinly sliced or shredded
80 g frozen edamame beans, deshelled, thawed
60 g red cabbage sliced or shredded
2 tbsp peanuts
2 tbsp chives

For the peanut sauce:
2 tbsp peanut butter
2 tbsp soy sauce
2 tsp sriracha sauce
2 tsp water
2 tsp honey

Method

1. Cook the noodles according to the package instructions, sieve and set aside.
2. Over a medium heat, add the oil to a frying pan and then the tofu. Season as you please.
3. Cook for about 5 minutes until the tofu pieces are browned then add the carrots and stir fry for a few minutes. Add the edamame beans until cooked (few minutes).
4. In a bowl, mix all the sauce ingredients well and feel free to add more water to create a thinner consistency.
5. Transfer the noodles into two dinner bowls, add the red cabbage as well as the tofu and vegetable mix on top.
6. Pour on some peanut sauce and sprinkle the fresh chives and peanuts on top.
7. Eat calmly and enjoy the dish to its fullest!

Baked salmon with cream cheese and vegetables
By Nutrition A-Z's clinical nutritionist, *Henriette Saevil*

Ingredients (serves 2)

2 medium carrots, chopped into quarters or cut into discs
60 g beetroot, peeled, scrubbed and chopped into small chunks
150 g frozen green beans, thawed
100 ml lactose-free milk or dairy-free milk of choice
100 g lactose-free cream cheese or natural yoghurt
Dried dill, pepper and salt to season
2 salmon fillets (2 x 125 g)

Method

1. Preheat the oven to 180°C/350°F (fan-assisted).
2. Add the vegetables into an oven-safe dish and allow them to pre-cook for about 10 minutes in the middle of the oven.
3. Whisk the milk and cream cheese together in a small cup and add any dried herbs of choice to the mixture.
4. Remove the skin from the salmon and cut the filets into smaller pieces; each filet can be cut into 10 pieces.
5. Take the vegetables out of the oven then place the salmon pieces on top.
6. Pour the milk and cream-cheese mixture over the salmon and vegetables.
7. Place the dish back into the middle of the oven and let it bake for about 15 minutes.
8. Remove the dish and serve.
9. Optional side dishes that go well include quinoa or brown rice. Bon appetit!

'Spanakorizo' Greek spinach-rice
By Nutrition A-Z's dietitian, 'Katia' Tassiou

This is a simple dish packed with Mediterranean flavours that serves as a great side dish to grilled or baked fish and can be topped with feta.

Ingredients (serves 5)

1 tbsp garlic infused oil
3 spring onions, green part only, finely chopped
700 g fresh spinach
1 handful fresh dill, chopped
1-2 tsp salt
1 tsp pepper
Juice of ½ lemon
180 g jasmine or basmati rice, rinsed under water in a strainer
    until the water running through it turns clear
30 g tomato paste
500 ml of water
2 tbsp olive oil

Method

1. In a wok-like or large-sized pan, add the garlic-infused oil and then sauté the spring onions over a medium heat until they soften and start to have a golden tint.
2. Add the spinach, dill, salt, pepper and lemon juice and stir until the spinach has wilted.
3. Then add the rice and tomato paste and stir for a couple of minutes.

4. Add the water to the pan and bring to a gentle boil. Then, turn the heat down and cover, letting the rice mixture simmer over a low heat covered for about 20-25 minutes until the water has been absorbed.

5. Once the rice is cooked and ready to go, transfer it to a serving dish, add 1 tbsp oil and additional lemon juice if you desire. Season with additional salt and pepper if needed as well as more fresh herbs such as dill and parsley.

For a low-chemical diet

A low-chemical diet is definitely one of the most challenging elimination diets to be on but the internationally recognised resource for all things FAILSAFE is the cookbook *Friendly Food* developed by the Royal Prince Alfred Hospital allergy unit in Sydney, Australia. As I've mentioned in the book, the RPAH elimination diet is the gold standard to-date that is used to diagnose food chemical intolerance. It is also known as FAILSAFE, which stands for 'free of additives and low in salicylates, amines and flavour enhancers'.

We have included a few of favourite low-chemical recipes, specifically low-histamine, that we created for this book.

Cottage cheese tzatziki (specifically low-histamine) By Nutrition A-Z's dietitian, 'Katia' Tassiou

Ingredients (serves 2)

200 g cottage cheese (choose the one with the fewest additives possible)
2 tbsp extra-virgin olive oil
2 garlic cloves
1 cucumber
1 handful fresh mint
1 handful fresh dill
1 tsp salt
Cracked black pepper to season

Method

1. Halve the cucumber lengthwise and use a small spoon to scoop out the seeds.
2. Grate the cucumber (no need to peel it).
3. Dice the mint and dill and mince the garlic.
4. Add the grated cucumber to a clean tea towel and wring it out over a sink to remove the excess liquid.
5. In a bowl, add all the ingredients and stir well to combine so the cucumber, herbs and garlic are well combined in the dip. Spoon into your serving bowl.
6. Serve your cottage cheese tzatziki with fresh crudites such as carrot sticks, celery and bell peppers or with our low-histamine crackers.

Rosemary sea-salt crackers (specifically low-histamine) By Nutrition A-Z's dietitian, 'Katia' Tassiou

Ingredients (serves 2)

90 g cassava flour
90 g arrowroot powder
½ tsp salt
½ tsp garlic powder
1 tsp onion powder
1 tbsp chopped fresh rosemary
80 ml extra-virgin olive oil
120 ml cold water
Salt and pepper for seasoning

Method

1. Preheat your oven to 200°C/390°F (fan-assisted).
2. In a medium mixing bowl, combine the cassava flour, arrow-root powder, salt, garlic powder, onion powder and rosemary.
3. Use a fork to whisk the oil and water into the flour mixture, forming a dough.
4. Roll the dough between two pieces of baking paper to form a thin rectangle.
5. Remove the top paper layer and slide the rolled-out dough along with the bottom piece of parchment onto a large baking tray.
6. Sprinkle generously with salt and pepper.
7. Bake for 20-25 minutes or until golden brown.
8. Allow it to cool before breaking into pieces for serving.

Roasted sweet pepper and cauliflower soup (specifically low-histamine)
By Nutrition A-Z's clinical nutritionist, *Henriette Saevil*

Ingredients (serves 2-3)

5 small-medium red peppers
1 small red onion
3 cloves garlic
2 tbsp coriander seeds
1.5 litres vegetable or chicken broth (there are numerous hista-mine-friendly broths on the market)
500 g cauliflower (1 medium head), chopped into bite-size florets
Optional protein choices: cooked chicken strips or chickpeas if tolerated

Spices:

1 small cayenne pepper

2 tsp ground sweet paprika

1 tsp salt

1 tbsp extra-virgin olive oil

Method

Step 1: Roast the peppers, garlic and onion

1. Preheat your oven to 180°C/250°F (fan-assisted).
2. Roughly chop the peppers and onion; peel the garlic.
3. In a bowl, mix the peppers, onion and garlic with olive oil, salt to taste and 1 tsp sweet paprika powder.
4. Spread the vegetables onto a baking tray and bake for 30-45 minutes or until they are lightly charred.

Step 2: Making the soup

1. Take a large pot, pour in the broth and add the cauliflower. Then allow it to simmer until the cauliflower is cooked (about 10-15 minutes).
2. Add the roasted vegetables and then remove the pot from your stove top.
3. Using an emulsion blender, blend the soup until it is smooth, then season with salt and 1 tsp sweet paprika powder. You can add lime juice and chilli if desired.
4. Serve topped with chopped coriander and, should you wish to add a protein of choice, this is best served with mixed in shredded chicken or chickpeas.

Bean and cabbage salad (FAILSAFE)

Ingredients (serves 4)

For the salad:
1 tbsp sunflower oil
80 g cashews, roughly chopped
1 small clove garlic, minced
270 g green beans, ends trimmed off and cut into smaller bites
500 g purple cabbage, shredded or finely chopped
400 g can borlotti beans, rinsed well

For the dressing:
200 g canned pears in juice (½ a can)
1 tsp citric acid
¼ cup sunflower oil
Salt to taste
1 small garlic clove, minced
1 tbsp water

Method

1. To make the dressing, add all the ingredients to a blender and purée until the desired consistency is achieved. You can thicken it by adding more pears; to thin it, add more water.
2. Place a pan over a medium heat and add the oil, then the cashews and garlic and cook until golden brown.
3. Add the green beans and cook until they are bright green and soft but still retain a crunch. Make sure you don't overcook them.
4. Once cooked, place in a bowl to cool down and start assembling your salad.

5. To a bowl, add the cabbage and borlotti beans, then mix in the green beans and cashews.

6. Add the dressing to your liking and mix your salad well. Add salt to taste.

7. This salad can be a lovely side-dish to your choice of protein such as firm tofu or fish.

My thank yous

I can already hear the 'wrap-it-up' Oscar's acceptance speech music in my head but first and foremost, there is no me without the two people who have shaped every bit of my being: my parents. Thank you for pushing us towards our dreams and passions without fear of failing. To my father, I wouldn't be making a difference in this field and breaking the taboo around poo if it weren't for you and the work that you do as an incredible gastroenterologist.

To my husband, thank you for being the voice of confidence I needed and for being the only person I can walk this life with. And to my little humans, thank you for allowing Mummy to write in the evenings only waking up with screams once I'd finished and made my way to bed. Well timed.

When it comes to mentorship, guidance and one of the very few people who understands the struggles and triumphs of my career, I cannot thank the one and only Maree Ferguson enough for literally being there from the start – from my very first clinical position at the PA back in Brisbane in 2008 to starting my own private practice years after on a different continent and more recently, this book. Thank you for being a constant support through it all.

To my fellow Birdhaus writers and Ana, thank you for creating the best accountability space to make the writing actually happen! Our fortnightly meetings behind our screens throughout the pandemic and sharing our writing journeys will always be a part of birthing this book.

To my Nutrition A-Z team and recipe contributors, Henriette and Katia, thank you for being the most valuable additions to a growing brand and for your kitchen creations that I can share with the world.

To Hammersmith Books and Georgina, thank you for taking a chance and allowing me to write about all things poo, bum and bloat in over 50,000 words. I cannot thank you enough for bringing this book to life and for offering the guidance I needed as I navigated this process, which was probably one of the hardest things I've ever had to do.

Amélie Buri, thank you for always producing the best illustrations and for creating THE ONE for this book's cover where it almost felt like you were in my head.

Michelle Sabatini, thank you for being the safe haven my mind has always needed throughout these years and for helping me navigate all the turbulences in my life by reminding me to breathe. I am forever grateful for all the work you've done in helping me understand Bob (context: we named my brain Bob).

My tribe of women and men who have been my cheerleaders throughout this journey, thank you. You all know who you are. Thank you for giving me the kick I needed whenever Bob plagued me with self-doubt and imposter syndrome. Thank you for the encouragement and for forcing me to celebrate all the little and big wins in my life. I am forever grateful to you all.

And finally, to you, my readers. Thank you for giving this book a spot on your bookshelf, nightstand or the bathroom floor (although fitting, please do reconsider its location). Jokes aside, YOU are the reason why I've dedicated the last few years to putting this book together and I truly hope I delivered. Thank you for pushing me to speak up and to break the taboo around topics that can truly disrupt our quality of life. A little bloat never killed nobody; a lot of uncomfortable bloat with dramatic changes in poo has.

References

Introduction

Mahan LK, Stump SE. Chapters 1, 26, 17, 28. In: *Krause's Food and Nutrition Therapy 12th Edition* Saunders (Elsevier); 2007.

Molnar A, Monroe H, Basri AH, et al. Tumors of the Digestive System: Comprehensive Review of Ancillary Testing and Biomarkers in the Era of Precision Medicine. *Current Oncology* 2023; 30(2): 2388-2404. https://doi.org/10.3390/curroncol30020182

WCRI. Cancer trends. World Cancer Research Fund International. www.wcrf.org/cancer-trends/ (accessed 28 March 2023)

Chapter 3: The one about your gut microbiome

Asnicar F, Berry SE, Valdes AM, Nguyen LH, et al. Microbiome connections with host metabolism and habitual diet from 1,098 deeply phenotyped individuals. *Nat Med* 2021; 27(2): 321-332.
doi: 10.1038/s41591-020-01183-8. PMID: 33432175; PMCID: PMC8353542.

Chen J, Chen X, Ho CL. Recent Development of Probiotic Bifidobacteria for Treating Human Diseases. *Front Bioeng Biotechnol* 2021; 9: 770248.
doi: 10.3389/fbioe.2021.770248. PMID: 35004640; PMCID: PMC8727868.

Dadar M, Tiwari R, Karthik K, Chakraborty S, Shahali Y, Dhama K. Candida albicans - Biology, molecular characterization, pathogenicity, and advances in diagnosis and control - An update. *Microb Pathog* 2018; 117: 128-138.
doi: 10.1016/j.micpath.2018.02.028. Epub 2018 Feb 16. PMID: 29454824.

Enaud R, Prevel R, Ciarlo E, Beaufils F, Wieërs G, Guery B, Delhaes L. The Gut-Lung Axis in Health and Respiratory Diseases: A Place for Inter-Organ and Inter-Kingdom Crosstalks. *Front Cell Infect Microbiol* 2020; 10: 9.
doi: 10.3389/fcimb.2020.00009. PMID: 32140452; PMCID: PMC7042389.

Ganesan K, Chung SK, Vanamala J, Xu B. Causal Relationship between Diet-Induced Gut Microbiota Changes and Diabetes: A Novel Strategy to Transplant Faecalibacterium prausnitzii in Preventing Diabetes. *Int J Mol Sci* 2018; 19(12): 3720. doi: 10.3390/ijms19123720. PMID: 30467295; PMCID: PMC6320976.

Hooper L, Martin N, Abdelhamid A, Davey Smith G. Reduction in saturated fat intake for cardiovascular disease. *Cochrane Database Syst Rev* 2015; (6): CD011737. doi: 10.1002/14651858.CD011737. Update in: Cochrane Database

Syst Rev. 2020 May 19;5:CD011737. PMID: 26068959.

Kho ZY, Lal SK. The Human Gut Microbiome: A Potential Controller of Wellness and Disease. *Front Microbiol* 2018; 9: 1835.
doi: 10.3389/fmicb.2018.01835. PMID: 30154767; PMCID: PMC6102370.

Kim KO, Gluck M. Fecal Microbiota Transplantation: An Update on Clinical Practice. *Clin Endosc* 2019; 52(2): 137-143.
doi: 10.5946/ce.2019.009. Epub 2019 Mar 26. PMID: 30909689; PMCID: PMC6453848.

Knezevic J, Starchl C, Tmava Berisha A, Amrein K. Thyroid-Gut-Axis: How Does the Microbiota Influence Thyroid Function? *Nutrients* 2020; 12(6): 1769.
doi: 10.3390/nu12061769. PMID: 32545596; PMCID: PMC7353203.

Salem I, Ramser A, Isham N, Ghannoum MA. The Gut Microbiome as a Major Regulator of the Gut-Skin Axis. *Front Microbiol* 2018; 9: 1459.
doi: 10.3389/fmicb.2018.01459. PMID: 30042740; PMCID: PMC6048199.

Valdes AM, Walter J, Segal E, Spector TD. Role of the gut microbiota in nutrition and health. *BMJ* 2018; 361: k2179.
doi: 10.1136/bmj.k2179. PMID: 29899036; PMCID: PMC6000740.

Chapter 4: The one on feeding your microbiome

Arnone D, Chabot C, Heba AC, Kökten T,et al. Sugars and Gastrointestinal Health. *Clin Gastroenterol Hepatol* 2022; 20(9): 1912-1924.e7.
doi: 10.1016/j.cgh.2021.12.011. PMID: 34902573.

Gill SK, Rossi M, Bajka B, Whelan K. Dietary fibre in gastrointestinal health and disease. *Nat Rev Gastroenterol Hepatol* 2021; 18(2): 101-116.
doi: 10.1038/s41575-020-00375-4. PMID: 33208922.

Kaur AP, Bhardwaj S, Dhanjal DS, Nepovimova E,et al. Plant Prebiotics and Their Role in the Amelioration of Diseases. *Biomolecules* 2021; 11(3): 440.
doi: 10.3390/biom11030440. PMID: 33809763; PMCID: PMC8002343.

Liu H, Wang J, He T, Becker S,et al. Butyrate: A Double-Edged Sword for Health? *Adv Nutr* 2018; 9(1): 21-29.
doi: 10.1093/advances/nmx009. PMID: 29438462; PMCID: PMC6333934.

Leeuwendaal NK, Stanton C, O'Toole PW, Beresford TP. Fermented Foods, Health and the Gut Microbiome. *Nutrients* 2022; 14(7): 1527.
doi: 10.3390/nu14071527. PMID: 35406140; PMCID: PMC9003261.

Lee E, Lee J. Impact of drinking alcohol on gut microbiota: recent perspectives on ethanol and alcoholic beverage. *Current Opinion in Food Science* 2021; 37: 91-97.

Minich DM. A Review of the Science of Colorful, Plant-Based Food and Practical Strategies for "Eating the Rainbow". *J Nutr Metab* 2019; 2019: 2125070.
doi: 10.1155/2019/2125070. Erratum in: *J Nutr Metab* 2020; 2020: 5631762. PMID: 33414957; PMCID: PMC7770496.

Ruiz-Ojeda, Francisco Javier et al. Effects of Sweeteners on the Gut Microbiota: A Review of Experimental Studies and Clinical Trials. *Advances in Nutrition* 2019;

10(S1): S31-S48. doi:10.1093/advances/nmy037

Schoeler M, Caesar R. Dietary lipids, gut microbiota and lipid metabolism. *Rev Endocr Metab Disord* 2019; 20(4): 461-472.
doi: 10.1007/s11154-019-09512-0. PMID: 31707624; PMCID: PMC6938793.

Wan Y, Wang F, Yuan J, et al. Effects of dietary fat on gut microbiota and faecal metabolites, and their relationship with cardiometabolic risk factors: a 6-month randomised controlled-feeding trial. *Gut* 2019; 0: 1-13.
doi: 10.1136/gutjnl-2018-317609.

Watson H, Mitra S, Croden FC, et al. A randomized trial of the effect of omega-3 polyunsaturated fatty acid supplements on the human intestinal microbiota. *Gut* 2018; 67(11): 1974-83. doi: 10.1136/gutjnl-2017-314968.

Wolters M, Ahrens J, Romaní-Pérez M, et al. Dietary fat, the gut microbiota, and metabolic health – A systematic review conducted within the MyNewGut project. *Clin Nutr* 2018; [vol? pages?]. doi: 10.1016/j.clnu.2018.12.024.

Chapter 5: The one on GORD – the volcanic throat

Bruno G, Zaccari P, Rocco G, Scalese G,et al. Proton pump inhibitors and dysbiosis: Current knowledge and aspects to be clarified. *World J Gastroenterol* 2019; 25(22): 2706-2719.
doi: 10.3748/wjg.v25.i22.2706. PMID: 31235994; PMCID: PMC6580352.

NICE (National Institute for Health and Care Excellence) Guidelines: Gastro-oesophageal reflux disease and dyspepsia in adults: investigation and management 2014. Updated 2019. www.nice.org.uk/guidance/cg184

Kang JH, Kang JY. Lifestyle measures in the management of gastro-oesophageal reflux disease: clinical and pathophysiological considerations. *Ther Adv Chronic Dis* 2015; 6(2): 51-64.
doi: 10.1177/2040622315569501. PMID: 25729556; PMCID: PMC4331235.

Sandhu DS, Fass R. Current Trends in the Management of Gastroesophageal Reflux Disease. *Gut Liver* 2018; 12(1): 7-16.
doi: 10.5009/gnl16615. PMID: 28427116; PMCID: PMC5753679.

Zohalinezhad ME, Imanieh MH, Samani SM, Mohagheghzadeh A,et al. Effects of Quince syrup on clinical symptoms of children with symptomatic gastroesophageal reflux disease: A double-blind randomized controlled clinical trial. *Complement Ther Clin Pract* 2015; 21(4): 268-276.
doi: 10.1016/j.ctcp.2015.09.005. Epub 2015 Sep 25. PMID: 26573454

Chapter 6: The one that's all about the bloat

Black CJ, Drossman DA, Talley NJ, Ruddy J, Ford AC. Functional gastrointestinal disorders: advances in understanding and management. *Lancet* 2020; 396(10263): 1664-1674.

doi: 10.1016/S0140-6736(20)32115-2. Epub 2020 Oct 10. PMID: 33049221.

Gibson PR, Shepherd SJ. Evidence-based dietary management of functional gastrointestinal symptoms: The FODMAP approach. *J Gastroenterol Hepatol* 2010; 25(2): 252-258. doi: 10.1111/j.1440-1746.2009.06149.x. PMID: 20136989.

Hungin APS, Mitchell CR, Whorwell P, Mulligan C, et al; European Society for Primary Care Gastroenterology. Systematic review: probiotics in the management of lower gastrointestinal symptoms - an updated evidence-based international consensus. *Aliment Pharmacol Ther* 2018; 47(8): 1054-1070.

doi: 10.1111/apt.14539. Epub 2018 Feb 20. PMID: 29460487; PMCID: PMC5900870.

Lacy BE, Cangemi D, Vazquez-Roque M. Management of Chronic Abdominal Distension and Bloating. *Clin Gastroenterol Hepatol* 2021; 19(2): 219-231.e1. doi: 10.1016/j.cgh.2020.03.056. Epub 2020 Apr 1. PMID: 32246999.

Lacy BE, Gabbard SL, Crowell MD. Pathophysiology, evaluation, and treatment of bloating: hope, hype, or hot air? *Gastroenterol Hepatol (NY)* 2011; 7(11): 729-739. PMID: 22298969; PMCID: PMC3264926.

Malagelada JR, Accarino A, Azpiroz F. Bloating and Abdominal Distension: Old Misconceptions and Current Knowledge. *Am J Gastroenterol* 2017; 112(8): 1221-1231. doi: 10.1038/ajg.2017.129. Epub 2017 May 16. PMID: 28508867.

Schmulson M, Chang L. Review article: the treatment of functional abdominal bloating and distension. *Aliment Pharmacol Ther* 2011; 33(10): 1071-1086. doi: 10.1111/j.1365-2036.2011.04637.x. PMID: 21488913.

Wilkinson JM, Cozine EW, Loftus CG. Gas, Bloating, and Belching: Approach to Evaluation and Management. *Am Fam Physician* 2019; 99(5): 301-309. PMID: 3081116.

Chapter 7: The one for the faecally challenged

Bristol Stool Scale Chart. Not dated. [cited 2017 Sept 1]. Available from: https://www.continence.org.au/pages/bristol-stool-chart.html

Gill SK, Rossi M, Bajka B, Whelan K. Dietary fibre in gastrointestinal health and disease. *Nat Rev Gastroenterol Hepatol* 2021; 18(2): 101-116. doi: 10.1038/s41575-020-00375-4. PMID: 33208922.

Hayat U, Dugum M, Garg S. Chronic constipation: Update on management. *Cleve Clin J Med* 2017; 84(5): 397-408. doi: 10.3949/ccjm.84a.15141. PMID: 28530898.

Hungin APS, Mitchell CR, Whorwell P, Mulligan C, et al, European Society for Primary Care Gastroenterology. Systematic review: probiotics in the management of lower gastrointestinal symptoms - an updated evidence-based international consensus. *Aliment Pharmacol Ther* 2018; 47(8): 1054-1070. doi: 10.1111/apt.14539. PMID: 29460487; PMCID: PMC5900870.

Jungersen M, Wind A, Johansen E, Christensen JE, Stuer-Lauridsen B, Eskesen D. The Science behind the Probiotic Strain Bifidobacterium animalis subsp. lactis BB-12(*). *Microorganisms* 2014; 2(2): 92-110.

doi: 10.3390/microorganisms2020092. PMID: 27682233; PMCID: PMC5029483.

References

Kapoor MP, Sugita M, Fukuzawa Y, Okubo T. Impact of partially hydrolyzed guar gum (PHGG) on constipation prevention: A systematic review and meta-analysis. *Journal of Functional Foods* 2017; 33: 52-66.
doi: 10.1016/j.jff.2017.03.028

Mearin F, Lacy BE, Chang L, Chey WD, Lembo AJ, Simren M, Spiller R. Bowel Disorders. *Gastroenterology* 2016: S0016-5085(16)00222-5.
doi: 10.1053/j.gastro.2016.02.031. PMID: 27144627.

Zółkiewicz J, Marzec A, Ruszczynski M, Feleszko W. Postbiotics-A Step Beyond Pre- and Probiotics. *Nutrients* 2020; 12(8): 2189.
doi: 10.3390/nu12082189. PMID: 32717965; PMCID: PMC7468815.

Chapter 8: The one on bumfluenza

Allen SJ, Martinez EG, Gregorio GV, Dans LF. Probiotics for treating acute infectious diarrhoea. *Cochrane Database Syst Rev* 2010; 2010(11): CD003048.
doi: 10.1002/14651858.CD003048.pub3. Update in: *Cochrane Database Syst Rev* 2020

Arasaradnam RP, Brown S, Forbes A, Fox MR, et al. Guidelines for the investigation of chronic diarrhoea in adults: British Society of Gastroenterology, 3rd edition. *Gut* 2018; 67(8): 1380-1399. doi: 10.1136/gutjnl-2017-315909.

Ciorba MA. A gastroenterologist's guide to probiotics. *Clin Gastroenterol Hepatol* 2012; 10(9): 960-968.
doi: 10.1016/j.cgh.2012.03.024. PMID: 22504002; PMCID: PMC3424311.

Hammer HF. Management of Chronic Diarrhea in Primary Care: The Gastroenterologists' Advice. *Dig Dis* 2021; 39(6): 615-621.
doi: 10.1159/000515219.

Hogan DE, Ivanina EA, Robbins DH. Probiotics: a review for clinical use. *Gastroenterology & Endoscopy News* 2018: 1-7. www.gastroendonews.com/Review-Articles/Article/05-21/Probiotics-forClinical-Use/63435

O'Brien L, Wall CL, Wilkinson TJ, Gearry RB. What Are the Pearls and Pitfalls of the Dietary Management for Chronic Diarrhoea? *Nutrients* 2021; 13(5): 1393.
doi: 10.3390/nu13051393.

Schiller LR, Pardi DS, Sellin JH. Chronic Diarrhea: Diagnosis and Management. *Clin Gastroenterol Hepatol* 2017; 15(2): 182-193.e3.
doi: 10.1016/j.cgh.2016.07.028.

Chapter 9: The one on irritable bowel syndrome (IBS)

Alammar N, Wang L, Saberi B, Nanavati J, Holtmann G, Shinohara RT, Mullin GE. The impact of peppermint oil on the irritable bowel syndrome: a meta-analysis of the pooled clinical data. *BMC Complement Altern Med* 2019; 19(1): 21.
doi: 10.1186/s12906-018-2409-0.

References

Black CJ, Ford AC. Best management of irritable bowel syndrome. *Frontline Gastroenterol* 2020; 12(4): 303-315. doi: 10.1136/flgastro-2019-101298.

Ducrotté P, Sawant P, Jayanthi V. Clinical trial: Lactobacillus plantarum 299v (DSM 9843) improves symptoms of irritable bowel syndrome. *World J Gastroenterol* 2012; 18(30): 4012-8. doi: 10.3748/wjg.v18.i30.4012.

Deiteren A, Camilleri M, Burton D, McKinzie S, Rao A, Zinsmeister AR. Effect of meal ingestion on ileocolonic and colonic transit in health and irritable bowel syndrome. *Dig Dis Sci* 2010; 55(2): 384-391. doi: 10.1007/s10620-009-1041-8.

Eskesen D, Jespersen L, Michelsen B, Whorwell PJ, Müller-Lissner S, Morberg CM. Effect of the probiotic strain Bifidobacterium animalis subsp. lactis, BB-12®, on defecation frequency in healthy subjects with low defecation frequency and abdominal discomfort: a randomised, double-blind, placebo-controlled, parallel-group trial. *Br J Nutr* 2015; 114(10): 1638-1646. doi: 10.1017/S0007114515003347.

Gibson PR, Shepherd SJ. Evidence-based dietary management of functional gastrointestinal symptoms: The FODMAP approach. *J Gastroenterol Hepatol* 2010; 25(2): 252-258. doi: 10.1111/j.1440-1746.2009.06149.x.

Jungersen M, Wind A, Johansen E, Christensen JE, Stuer-Lauridsen B, Eskesen D. The Science behind the Probiotic Strain *Bifidobacterium animalis* subsp. *lactis* BB-12®. *Microorganisms* 2014; 2(2):92-110. https://doi.org/10.3390/microorganisms2020092

Nishida S, Gotou M, Akutsu S, Ono M, et al. Effect of Yoghurt Containing *Bifidobacterium lactis* BB-12® on Improvement of Defecation and Fecal Microflora of Healthy Female Adults. *Milk Science* 2004; 53(2): 71-80.

Nobaek S, Johansson ML, Molin G, Ahrné S, Jeppsson B. Alteration of intestinal microflora is associated with reduction in abdominal bloating and pain in patients with irritable bowel syndrome. *Am J Gastroenterol* 2000; 95(5): 1231-1238. doi: 10.1111/j.1572-0241.2000.02015.x.

Patel A, Hasak S, Cassell B, Ciorba MA, et al. Effects of disturbed sleep on gastrointestinal and somatic pain symptoms in irritable bowel syndrome. *Aliment Pharmacol Ther* 2016; 44(3): 246-258. doi: 10.1111/apt.13677.

Zulman DM, Haverfield MC, Shaw JG, Brown-Johnson CG, et al. Practices to Foster Physician Presence and Connection With Patients in the Clinical Encounter. *JAMA* 2020; 323(1): 70-81. doi: 10.1001/jama.2019.19003..

Chapter 10: The one that's all about food intolerance

Dale HF, Biesiekierski JR, Lied GA. Non-coeliac gluten sensitivity and the spectrum of gluten-related disorders: an updated overview. *Nutr Res Rev* 2019; 32(1): 28-37. doi: 10.1017/S095442241800015X.

Dev S, Mizuguchi H, Das AK, Matsushita C, et al. Suppression of histamine signaling by probiotic Lac-B: a possible mechanism of its anti-allergic effect. *J Pharmacol Sci* 2008; 107(2): 159-166. doi: 10.1254/jphs.08028fp

References

Gargano D, Appanna R, Santonicola A, De Bartolomeis F, et al. Food Allergy and Intolerance: A Narrative Review on Nutritional Concerns. *Nutrients* 2021; 13(5): 1638. doi: 10.3390/nu13051638.

Kung HF, Lee YC, Huang YL, Huang YR, Su YC, Tsai YH. Degradation of histamine by Lactobacillus plantarum isolated from Miso products. *J Food Prot* 2017; 80(10): 1682-1688.

Oksaharju A, Kankainen M, Kekkonen RA, Lindstedt KA, et al. Probiotic Lactobacillus rhamnosus downregulates FCER1 and HRH4 expression in human mast cells. *World J Gastroenterol* 2011; 17(6): 750-759. doi: 10.3748/wjg.v17.i6.750.

Sánchez-Pérez S, Comas-Basté O, Veciana-Nogués MT, Latorre-Moratalla ML, Vidal-Carou MC. Low-Histamine Diets: Is the Exclusion of Foods Justified by Their Histamine Content? *Nutrients* 2021; 13(5): 1395. doi: 10.3390/nu13051395.

Swain A, Soutter V, Loblay R. RPAH Elimination Diet. *Handbook*: Allergy unit Royal Prince Alfred Hospital; Allergy Unit, Royal Prince Alfred Hospital: Sydney, Australia, 2009.

Tuck CJ, Biesiekierski JR, Schmid-Grendelmeier P, Pohl D. Food Intolerances. *Nutrients* 2019; 11(7): 1684. doi: 10.3390/nu11071684

Chapter 11: The one on bum-dumplings – haemorrhoids

Greenwald DA, Choudhary C, Levine J. Common disorders of the anus and rectum. *American College of Gastroenterology* 2018. gi.org/topics/hemorrhoids-and-other-anal-disorders/

McKeown NM, Fahey GC Jr, Slavin J, van der Kamp JW. Fibre intake for optimal health: how can healthcare professionals support people to reach dietary recommendations? *BMJ* 2022; 378: e054370. doi: 10.1136/bmj-2020-054370. PMID: 35858693; PMCID: PMC9298262.

Perera N, Liolitsa D, Iype S, Croxford A, et al. Phlebotonics for haemorrhoids. *Cochrane Database Syst Rev* 2012; (8): CD004322. doi: 10.1002/14651858.CD004322.pub3. PMID: 22895941

Sandler RS, Peery AF. Rethinking What We Know About Hemorrhoids. *Clin Gastroenterol Hepatol* 2019; 17(1): 8-15. doi: 10.1016/j.cgh.2018.03.020. Epub 2018 Mar 27. PMID: 29601902; PMCID: PMC7075634.

Chapter 12: The one that ends with a blissful gut

Alammar N, Wang L, Saberi B, Nanavati J, Holtmann G, Shinohara RT, Mullin GE. The impact of peppermint oil on the irritable bowel syndrome: a meta-analysis of the pooled clinical data. *BMC Complement Altern Med* 2019; 19(1): 21.

References

doi: 10.1186/s12906-018-2409-0.

Black CJ, Ford AC. Best management of irritable bowel syndrome. *Frontline Gastroenterol* 2020; 12(4): 303-315. doi: 10.1136/flgastro-2019-101298.

Camfield DA, Stough C, Farrimond J, Scholey AB. Acute effects of tea constituents L-theanine, caffeine, and epigallocatechin gallate on cognitive function and mood: a systematic review and meta-analysis. *Nutr Rev* 2014; 72(8): 507-522. doi: 10.1111/nure.12120. Epub 2014 Jun 19. PMID: 24946991.

D'Afflitto M, Upadhyaya A, Green A, Peiris M. Association Between Sex Hormone Levels and Gut Microbiota Composition and Diversity-A Systematic Review. *J Clin Gastroenterol* 2022; 56(5): 384-392.
doi: 10.1097/MCG.0000000000001676. PMID: 35283442; PMCID: PMC7612624.

Deiteren A, Camilleri M, Burton D, McKinzie S, Rao A, Zinsmeister AR. Effect of meal ingestion on ileocolonic and colonic transit in health and irritable bowel syndrome. *Dig Dis Sci* 2010; 55(2): 384-391. doi: 10.1007/s10620-009-1041-8.

Ducrotté P, Sawant P, Jayanthi V. Clinical trial: Lactobacillus plantarum 299v (DSM 9843) improves symptoms of irritable bowel syndrome. *World J Gastroenterol* 2012; 18(30): 4012-8. doi: 10.3748/wjg.v18.i30.4012.

Eskesen D, Jespersen L, Michelsen B, Whorwell PJ, Müller-Lissner S, Morberg CM. Effect of the probiotic strain Bifidobacterium animalis subsp. lactis, BB-12®, on defecation frequency in healthy subjects with low defecation frequency and abdominal discomfort: a randomised, double-blind, placebo-controlled, parallel-group trial. *Br J Nutr* 2015; 114(10): 1638-1646.
doi: 10.1017/S0007114515003347.

Fowler S, Hoedt EC, Talley NJ, Keely S, Burns GL. Circadian Rhythms and Melatonin Metabolism in Patients With Disorders of Gut-Brain Interactions. *Front Neurosci* 2022; 16: 825246. doi: 10.3389/fnins.2022.825246. PMID: 35356051; PMCID: PMC8959415.

GBD 2015 Mortality and Causes of Death Collaborators. Global, regional, and national life expectancy, all-cause mortality, and cause-specific mortality for 249 causes of death, 1980–2015: a systematic analysis for the Global Burden of Disease Study 2015. *Lancet* 2016; 388: 1459-1544.
https://doi.org/10.1016/S0140-6736(16)31012-1

Gibson PR, Shepherd SJ. Evidence-based dietary management of functional gastrointestinal symptoms: The FODMAP approach. *J Gastroenterol Hepatol* 2010; 25(2): 252-258. doi: 10.1111/j.1440-1746.2009.06149.x.

Hamasaki H. Effects of Diaphragmatic Breathing on Health: A Narrative Review. *Medicines (Basel)* 2020; 7(10): 65.
doi: 10.3390/medicines7100065. PMID: 33076360; PMCID: PMC7602530.

Hopper SI, Murray SL, Ferrara LR, Singleton JK. Effectiveness of diaphragmatic breathing for reducing physiological and psychological stress in adults: a quantitative systematic review. *JBI Database System Rev Implement Rep* 2019; 17(9): 1855-1876. doi: 10.11124/JBISRIR-2017-003848. PMID: 31436595.

Jungersen M, Wind A, Johansen E, Christensen JE, Stuer-Lauridsen B, Eskesen D. The Science behind the Probiotic Strain *Bifidobacterium animalis* subsp. *lactis*

References

BB-12®. *Microorganisms* 2014; 2(2): 92-110.
https://doi.org/10.3390/microorganisms2020092

Kwa M, Plottel CS, Blaser MJ, Adams S. The Intestinal Microbiome and Estrogen
Receptor-Positive Female Breast Cancer. *J Natl Cancer Inst* 2016; 108(8): djw029.
doi: 10.1093/jnci/djw029. PMID: 27107051; PMCID: PMC5017946.

Meleine M, Matricon J. Gender-related differences in irritable bowel syndrome:
Potential mechanisms of sex hormones. World Journal of Gastroenterology
2014; 20(22): 6725-6743.
doi: 10.3748/wjg.v20.i22.6725. PMID: 24944465; PMCID: PMC4051914.

Nobaek S, Johansson ML, Molin G, Ahrné S, Jeppsson B. Alteration of intestinal
microflora is associated with reduction in abdominal bloating and pain in
patients with irritable bowel syndrome. *Am J Gastroenterol* 2000; 95(5): 1231-
1238. doi: 10.1111/j.1572-0241.2000.02015.x.

Patel A, Hasak S, Cassell B, Ciorba MA, et al. Effects of disturbed sleep on
gastrointestinal and somatic pain symptoms in irritable bowel syndrome.
Aliment Pharmacol Ther 2016; 44(3): 246-258. doi: 10.1111/apt.13677.

Rogers M, Coates AM, Banks S. Meal timing, sleep, and cardiometabolic outcomes.
Current Opinion in Endocrine and Metabolic Research 2021; 18: 128-132.
https://doi.org/10.1016/j.coemr.2021.03.006.

Schumann D, Langhorst J, Dobos G, Cramer H. Randomised clinical trial: yoga
vs a low-FODMAP diet in patients with irritable bowel syndrome. *Aliment
Pharmacol Ther* 2018; 47(2): 203-211. doi: 10.1111/apt.14400

Selhub EM, Logan AC, Bested AC. Fermented foods, microbiota, and mental
health: ancient practice meets nutritional psychiatry. *J Physiol Anthropol* 2014;
33(1): 2. doi: 10.1186/1880-6805-33-2. PMID: 24422720; PMCID: PMC3904694.

Smith RP, Easson C, Lyle SM, Kapoor R, et al. Gut microbiome diversity is
associated with sleep physiology in humans. *PLoS One* 2019; 14(10): e0222394.
doi: 10.1371/journal.pone.0222394. PMID: 31589627; PMCID: PMC6779243.

Trzeciak P, Herbet M. Role of the Intestinal Microbiome, Intestinal Barrier and
Psychobiotics in Depression. *Nutrients* 2021; 13(3): 927.
doi: 10.3390/nu13030927. PMID: 33809367; PMCID: PMC8000572.

Turjeman S, Collado MC, Koren O. The gut microbiome in pregnancy and
pregnancy complications. *Current Opinion in Endocrine and Metabolic
Research* 2021; 18: 133-138. https://doi.org/10.1016/j.coemr.2021.03.004

Watson NF, Badr MS, Belenky G, Bliwise DL, et al. Recommended Amount of Sleep
for a Healthy Adult: A Joint Consensus Statement of the American Academy of
Sleep Medicine and Sleep Research Society. *Sleep* 2015; 38(6): 843-844.
doi: 10.5665/sleep.4716. PMID: 26039963; PMCID: PMC4434546.

Waxenbaum JA, Reddy V, Varacallo M. Anatomy, Autonomic Nervous System.
[Updated 2021 Jul 29]. In: StatPearls [Internet]. Treasure Island (FL): StatPearls
Publishing; 2022

Zulman DM, Haverfield MC, Shaw JG, Brown-Johnson CG, et al. Practices to Foster
Physician Presence and Connection With Patients in the Clinical Encounter.
JAMA 2020; 323(1): 70-81. doi: 10.1001/jama.2019.19003.

Glossary

Acid reflux The movement of acid from the stomach backward into the oesophagus due to the weakening or relaxation of the lower oesophageal sphincter or LES (a circular muscle that separates your oesophagus from your stomach).

Anal fissures are tears along the tissue lining of the anus. They can occur if you pass dry, hard stool (i.e. if you are constipated) but also during childbirth or anal intercourse.

Artificial sweeteners are sugar-substitutes that are created in the lab and mimic the taste of sugar without the calories. Examples of these are acesulfame K, aspartame, cyclamate, saccharin and sucralose. These sweeteners are typically found in processed foods such as baked goods, soft drinks, canned foods, sweets, jams, dairy products and basically most products labelled as 'high-protein, low-carb.'

Bile acid malabsorption (BAM) Also known as bile acid diarrhoea, BAM is when bile acids do not get reabsorbed in the small intestine and instead continue on to the colon, causing irritation and excess water secretions that lead to loose stool.

Bristol Stool Chart An internationally recognised chart that describes the different types and shapes of stool and provides a useful standard reference.

Dysbiosis describes an unfavourable imbalance of our gut microbes. Dysbiosis is considered to be an unfavourable state that may involve a reduction in microbial diversity, changes in the types of beneficial species, or changes in how beneficial microbes function.

Enemas vs colonics An enema is a procedure where a tube is inserted into the rectum via the anus and liquid is injected (normally water) to stimulate the bowels to fully empty. It is generally used prior to surgery if indicated or for the occasional relief of severe constipation. A colonic or colonic irrigation is claimed to cleanse the bowel, ridding it of toxins. The difference between an enema and colonic irrigation is the amount of fluid being flushed up the intestine – with an enema, we are looking at a one-off infusion whereas, with a colonic, it is continuous i.e. multiple infusions for about 45 minutes.

FODMAPs is a term given to a group of poorly digested, rapidly fermentable carbohydrate food molecules. The term FODMAPs stands for 'fermentable oligo-, di- and mono-saccharides and polyols'.

Fructans, also known as oligosaccharides, are a type of fermentable sugars found in wheat, rye and some vegetables.

Galactans, also known as galacto-oligosaccharides, are the fermentable sugars found in legumes such as beans, lentils, chickpeas and soya.

Gastrocolic reflex is your body's response to consuming food.

GORD (aka GERD – gastro-(o)esophageal reflux disease) is a collection of symptoms due to the reflux of gastric contents such as acid into the oesophagus, larynx, mouth and/or lungs.

Haemorrhoids (aka 'piles' and 'bum dumplings') are swollen veins in the anus and lower rectum. There are four grades:
    Grade I: No signs of prolapse
    Grade II: Prolapsed but retract on their own

Grade III: Prolapsed but can be manually pushed back in

Grade IV: Prolapsed but cannot be pushed back in without a lot of pain.

Natural no-sugar sweeteners include the popular stevia. Funnily enough, the commercial stevia that you end up buying off the shelves barely contains any whole-leaf stevia at all, putting the term 'natural' up for question. Such products are made from a highly processed stevia leaf extract called Reb-A so unless you're growing stevia at home, there's not much natural left. Stevia is popular amongst a number of hipster-looking brands that sell iced teas, protein bars and shakes, and bakeries claiming to look after your health.

Nocebo effect is the negative placebo effect when a person is already expecting to experience unpleasant symptoms going into a challenge and has a biased expectation about the consequences.

Non-coeliac wheat sensitivity (NCWS) is a condition that refers to the intolerance to wheat. NCWS is a recent entity that was born out of a small percentage of people who did not fall under the category of being sufferers of coeliac disease or a wheat allergy.

Oestrobolome is the name for a specific collection of gut microbes that are capable of regulating the metabolism of oestrogen.

Oxidative stress is the result of an imbalance between the production of reactive oxygen species (free radicals) that result from our metabolic processes and our antioxidant defences.

Phlebotonics are a class of medications of plant origin that are believed to improve the health and strength of blood vessel walls.

Polyols are fermentable sugar alcohols that are naturally found in some fruits and vegetables but can also be manufactured in labs and used as sweeteners added to foods.

Polyphenols are nutrients found in many plant-based foods and drinks and are responsible for their vibrant colours. They are known antioxidants, protecting our body's cells from damage caused by harmful molecules called free radicals. They are also known to have prebiotic properties.

Postbiotics are beneficial bioactive compounds produced by our gut microbes during the fermentation process.

Prebiotics are types of fibre that are food for the good bacteria (i.e. probiotics) in the gut.

Probiotics are live microorganisms that, when consumed in adequate amounts, confer a health benefit on the host (us!).

Psychobiotics are probiotic bacteria that, when ingested in appropriate quantities, exert a positive health benefit on mental wellbeing including psychiatric illnesses.

Runner's gut is the term used for loose stools brought on by running/high-energy movement.

SIBO (small intestinal bacterial overgrowth) is having unhealthily high levels of microbes in the small intestine.

Skin tags A skin tag that forms around the anus is just the extra skin left behind after a blood clot in an external haemorrhoid has resolved.

Sugar alcohols are carbohydrates that are naturally found in some fruit and vegetables but can also be manufactured in a lab. They do contain calories but still in lower quantities than table sugar, hence their being a popular alternative. Sugar alcohols are found in chocolate, lollies, chewing gum and even toothpaste. No, they do not contain any ethanol despite the 'alcohol' part. Sugar alcohols are easy to identify as they end in '-ol' and include sorbitol, mannitol, maltitol, lactitol, erythritol and xylitol. You'll notice that foods containing sugar alcohols come with a warning that 'excess consumption may have a laxative effect'. Given that they travel all the way to our large bowel and are poorly absorbed there, excess consumption can cause bloating, stomach pain, gas and diarrhoea.

Thrombosed haemorrhoid This is when your haemorrhoid develops a blood clot in a haemorrhoidal vein, causing an obstruction in blood flow. Thrombosed haemorrhoids can be very painful and swollen and may cause rectal bleeding.

Vagus nerve is the longest nerve in the body ('vagus' means 'wandering') and consists of a network of nerves central to the body's autonomic nervous system that controls involuntary functions.

Index

abdominal bloating *see* bloating

acesulfame K, 58

acid (stomach)
  ageing and, 11
  increased pressure on, 65
  increased secretion, 65
  reflux *see* gastro-oesophageal
    reflux disease; reflux

actinidin protease, 106

activated charcoal, 90

additives (food), 169

ageing, 11–12
  food intolerance and, 12,
    168–169

Akkermansia muciniphila, 57

alcohol, 55–57
  IBS and, 140
  in kombucha, 48

allergy, *see* food allergy

alternative therapy, GORD, 71–72

amines, 167, 174, 175

amino acids, 38

amylase trypsin inhibitors (ATIs),
  171, 172

animal (and animal-based) foods,
  52–53, 54

antacids, 70

antibiotics, 115
  Clostridium difficile and, 28–29

diarrhoea associated with,
  28–29, 45, 115, 128

diarrhoea (chronic) treated
  with, 123

in small intestinal bacterial
  overgrowth, 121

antibodies (immunoglobulins)
  IgE, 166, 167
  IgG, 169, 170

antidepressants
  IBS, 137
  probiotics as supplements to,
    234

antispasmodics, 137, 138

anus, 9–10, 186
  fissures, 186, **285**
  skin tags, 186, **288**

anxiety, 137, 156–159
  author's, 216
  IBS and, 156–159

arabynoxylan, 42

artificial colours, 175

artificial sweeteners, 58–60, 82–83,
  285
  bloating and, 82–83
  in kombucha, 48

aspartame, 58

autoimmune disease/disorders,
  22–23, 118, 172, 174, 205

Note: bold indicates glossary definitions.

Index

Note: bold indicates glossary definitions.

Index

Note: bold indicates glossary definitions.

Note: bold indicates glossary definitions.

Note: bold indicates glossary definitions.

Note: bold indicates glossary definitions.

Note: bold indicates glossary definitions.

Note: bold indicates glossary definitions.

Also from Hammersmith Health Books...

Plant-Based Nutrition in Clinical Practice

Edited by Dr Shireen Kassam, Dr Zahra Kassam
and Lisa Simon RD

with a team of 27 contributing specialist authors

In *Plant-Based Nutrition in Clinical Practice*, a team of specialist
authors, edited by two hospital doctors and registered dietitian
Lisa Simon, curate the current knowledge base that supports
a whole food, plant-based diet as the best way to address both
individual and planetary health. Taking both a holistic and
systems-based approach, the book presents the uses, benefits
and practical application of a plant-based diet to support
patient care, disease prevention and management.

Also from Hammersmith Health Books...

Eating Plant-Based

Scientific Answers to your Nutrition Questions

By Dr Shireen Kassam and Dr Zahra Kassam

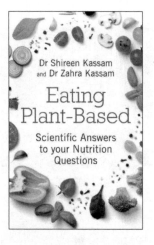

Following a Q&A format, Shireen and Zahra Kassam, both medical doctors specialising in cancer treatment, answer all the commonly asked questions you may be concerned about when considering transitioning to a plant-based diet. Simple and straightforward answers are supported with the scientific background and references to the latest published research findings.